CHAPMAN & HALL/CRC FINANCIAL MATHEMATICS SERIES

Robust Libor Modelling and Pricing of Derivative Products

CHAPMAN & HALL/CRC
Financial Mathematics Series

Aims and scope:
The field of financial mathematics forms an ever-expanding slice of the financial sector. This series aims to capture new developments and summarize what is known over the whole spectrum of this field. It will include a broad range of textbooks, reference works and handbooks that are meant to appeal to both academics and practitioners. The inclusion of numerical code and concrete real-world examples is highly encouraged.

Series Editors

M.A.H. Dempster
Centre for Financial Research
Judge Institute of Management
University of Cambridge

Dilip B. Madan
Robert H. Smith School of Business
University of Maryland

Rama Cont
Centre de Mathematiques Appliquees
Ecole Polytechnique

Proposals for the series should be submitted to one of the series editors above or directly to:
CRC Press UK
23 Blades Court
Deodar Road
London SW15 2NU
UK

CHAPMAN & HALL/CRC FINANCIAL MATHEMATICS SERIES

Robust Libor Modelling and Pricing of Derivative Products

HG
6024.5
S36
2005
WEB

John Schoenmakers

Chapman & Hall/CRC
Taylor & Francis Group
Boca Raton   London   New York   Singapore

Published in 2005 by
Chapman & Hall/CRC
Taylor & Francis Group
6000 Broken Sound Parkway NW, Suite 300
Boca Raton, FL 33487-2742

© 2005 by Taylor & Francis Group, LLC
Chapman & Hall/CRC is an imprint of Taylor & Francis Group

No claim to original U.S. Government works
Printed in the United States of America on acid-free paper
10 9 8 7 6 5 4 3 2 1

International Standard Book Number-10: 1-58488-441-X (Hardcover)
International Standard Book Number-13: 978-1-5848-8441-5 (Hardcover)
Library of Congress Card Number 2004065976

Library of Congress Cataloging-in-Publication Data

Schoenmakers, John.
   Robust Libor modelling and pricing of derivative products / John Schoenmakers.
      p. cm.
   Includes bibliographical references and index.
   ISBN 1-58488-441-X (alk. paper)
    1. Interest rate futures—Mathematical models. 2. Interest rates—Mathematical models.
  3. Derivative securities—Prices—Mathematical models. I. Title.
   HG6024.5.S36 2005
   332.64'57'0151--dc22   2004065976

Taylor & Francis Group
is the Academic Division of T&F Informa plc.

Visit the Taylor & Francis Web site at
http://www.taylorandfrancis.com

and the CRC Press Web site at
http://www.crcpress.com

Veur Barbara, Luc, Tim

Preface

This book may be considered, on the one hand, as a fast and handy introduction to the area of (Libor) interest rate modelling and pricing, intended for students in finance, junior financial engineers, and other interested people with a mathematical background, which covers some knowledge of basic probability theory and stochastic processes.[1] On the other hand, due to an innovative treatment of issues concerning Libor calibration and pricing of exotic instruments, including new valuation procedures for products with Bermudan style exercise features, the book is suited particularly to the more experienced practitioner and the researcher in this field as well.

Since its development back in the mid nineties the Libor Market Model is still one of the most popular and advanced tools for modelling interest rates and interest rate derivatives at the so called fixed income desks. Nevertheless, a useful procedure for "implied" calibration of the model, which avoids historical or subjective information but exclusively relies on market prices of liquidly traded products, has been a perennial problem. In recent studies, however, we[2] tackled this problem and developed fast and robust implied methods for calibrating Libor. This calibration approach, its impact on Libor derivative pricing, and the development of new methods for pricing complicated exotic (Bermudan) structures in the Libor model provided the principle motivation for writing this book.

In fact, several key ideas and results presented in this work were obtained during our research and consulting on Libor modelling and derivative pricing over the last years. Practically right after the development and appearance of the Libor market model in early versions of the now broadly known standard papers of Brace, Gatarek & Musiela, Jamshidian, and Miltersen, Sandman & Sonderman, we got involved in consulting projects on implementation and application of this new promising model for pricing exotic interest rate products at fixed income desks. Naturally, the first job was to construct stable methods for calibration of the model to liquidly traded caps and swaptions. However, soon it turned out that in spite of many nice properties of the Libor market model, calibration of a more factor model (i.e., more than one driving random source) tends to suffer from stability problems. In particular, difficulties of intrinsic nature came out by the determination of the trade off between the explanation power of forward Libor volatilities versus forward Libor correlations from market data, even if these quantities were stuck into a very parsimonious framework. Loosely speaking, even in a parsimonious setting it turned out possible to match the same market quotes with essentially different calibrations! In the present work we present a remedy for this problem via incorporation of a collateral swaption market criterion in the cal-

---

[1] The reader is recommended to consult some standard texts on this subject if necessary.

[2] Throughout the text "we" stands for the present author and his circle of research associates.

ibration procedure. Basically, via this collateral market concept stability is retained in the calibration. Indeed, for a parsimonious model the behaviour of the model parameters implied by the market via a thus enhanced calibration procedure proved to be very regular when going from one calibration date to another, similar to implied Black-Scholes volatilities observed from standard stock options!

In the area of Bermudan style exotic Libor products we expose effective standard methods such as the Andersen method for constructing lower bounds and the Rogers method for constructing (dual) upper bounds. Moreover, we introduce a recently developed generic iteration procedure (Kolodko & Schoenmakers 2004, Bender & Schoenmakers 2004) for the valuation of standard Bermudan and multiple callable Bermudan instruments. A very appealing feature of the presented iteration procedure is the fact that it applies to any underlying model because one only needs to simulate the underlying process just like in the (naïve) backward dynamic program. In contrast to the backward dynamic program, however, the degree of nesting of simulations does not grow with the number of exercise dates, but can be chosen in advance. Usually, a quadratic Monte Carlo simulation gives extremely good results.

The robust implied calibration approach has been implemented together with pricing routines for various exotic products such as trigger swaps, flexi caps and Bermudan swaptions. Using this software, realised as an Excel add-in, we illustrate in various practical cases the calibration procedure and its impact on exotics pricing.

The material is embedded in a compact and self contained review of the necessary financial mathematical theory on interest rate modelling given in Chapter 1.

Chapter 2 concentrates on economically sensible parametrisations of the Libor Market Model. Particularly, the presented low parameter correlation structures in this chapter may be considered as new.

In Chapter 3 we expose an intrinsic stability problem connected with direct least squares calibration methods, supported by both practical examples, and examples generated in our "laboratory". We here introduce a stable calibration procedure by incorporating a criterion related to the swaption market.

Chapter 4 deals with Libor exotics pricing based on real life calibrated models and we here consider the impact of different calibrations on a variety of popular exotic derivative structures. For instance, we show that certain instruments (e.g., the ratchet cap) are quite sensitive for the number of driving factors in the model.

Our new developments on Bermudan style derivative pricing are presented in Chapter 5. In this chapter we also explain the Andersen (1999) method, dual upper bound methods due to Rogers, Haugh & Kogan, and Jamshidian, and introduce an effective algorithm for constructing dual upper bounds.

In Chapter 6 we deal with suitable lognormal approximations for the Libor model and show that in comparison with standard Euler simulation of the Libor dynamics, it is more efficient to simulate prices for long dated instruments using such approximations.

We think it is fair to say that the thorny stability issues concerning (implied) calibration of Libor models, and its consequences for derivative pricing, are not well emphasized and discussed in the literature so far. Moreover, particularly on the subject of Bermudan style derivatives, several new ideas which may serve as a starting point for further research are presented.

Throughout the book I (the author) strove for clarity at every point of issue, and supporting examples and illustrations are given where possible. A trial version of the Excel add-in, used for the empirical experiments, is available on request.

Acknowledgments. This book was composed at the Weierstrass Institute for Applied Analysis and Stochastics in Berlin and I (the author) thank the institute, and particularly my direct colleagues, for the pleasant working environment. Much of the new material presented in this book is developed in cooperation with the DFG research center MATHEON, *Mathematics for key technologies*, in Berlin which deserves my thanks as well. A special thanks goes to my former colleague Oliver Reiß for valuable discussions and developing together with me the C++/Excel prototype implementation. Further I would like to thank, at the Weierstrass Institute, Christian Bender and Anastasia Kolodko for many useful remarks. This work also strongly benefited from cooperation with many persons outside WIAS, among others I mention Brian Coffey, Arnold Heemink, Hermann Haaf, Farshid Jamshidian, Jörg Kienitz, Pieter Klaassen, Peter Kloeden, Christoph März, Grigori Milstein, Geert Jan Olsder, Thorsten Sauder, Samuel Schwalm, and Martin Schweizer.

John Schoenmakers
Berlin

Contents

xii

List of Tables

List of Figures

Symbol Description

$\emptyset$	the empty set	$\mathbb{R}_{++}$	the set $\{x \in \mathbb{R} : x > 0\}$
$A \cup B$	the union of sets A and B	$\mathbb{R}^m$	the m-dimensional Euclid-
$A \cap B$	the intersection of sets A and B		ean space; $\mathbb{R}^1 = \mathbb{R}$
		$:=$	defined as or denoted by
$a \in A$	a belongs to the set A	$\equiv$	identically equal to
$a \notin A$	a belongs not to the set A	$\approx$	approximately equal to
$\mathbb{N}$	The set $\{1,2,3,\dots\}$	$\sim$	distributed as
$\mathbb{N}_0$	The set $\{0,1,2,3,\dots\}$	(a,b)	the open interval $a < x < b$
$\mathbb{R}$	the set of real numbers		in $\mathbb{R}$
$\mathbb{R}_+$	the set $\{x \in \mathbb{R} : x \geq 0\}$	$[a,b]$	the closed interval

$a \leq x \leq b$ in $\mathbb{R}$

$\max(a, b)$ maximum of two real numbers a and b

$\min(a, b)$ minimum of two real numbers a and b

$\inf A$ the infimum of a set A

$\sup A$ the supremum of a set A

$O(t^q)$, $t \to 0$ or ∞
expression divided by t^q is bounded for $t \to 0$ or ∞, resp.

$o(t^q)$, $t \to 0$ or ∞
expression divided by t^q converges to zero for $t \to 0$ or ∞, resp.

$a \vee b$ logical a "or" b for booleans a and b, $\max(a, b)$ for real numbers a and b

$a \wedge b$ logical a "and" b for booleans a and b, $\min(a, b)$ for real numbers a and b

a^+ $\max(a, 0)$

a^- $-\min(a, 0)$

$1_A(x)$ for a set A equals 1 when $x \in A$, 0 when $x \notin A$

1_P for a boolean P equals 1 when P is true, 0 when P is false

$A^\top$ the transpose of the matrix (or vector) A

$|x|$ norm $\sqrt{x^\top x}$ when $x \in \mathbb{R}^m$, $m \geq 1$.

$\text{span}\{v_1, \ldots, v_k\}$
the set of linear combinations $\sum_{i=1}^{k} \alpha_i v_i$ of elements v_i of a linear space

$\text{rank}(A)$ the dimension of the column space of a matrix A

EX the mathematical expectation of a random variable X

$E^{\mathcal{F}} X$ the conditional expectation X given the σ-algebra $\mathcal{F}$

a.s. almost surely

Var Variance

Cov Covariance

Cor Correlation

Chapter 1

Arbitrage-Free Modelling of Effective Interest Rates

In this first chapter we provide a general framework for modelling of effective forward interest rates such as forward Libor and swap rates. Throughout the whole book Libor may stand for LIBOR (London Inter Bank Offer Rate), EurIBOR (European Inter Bank Offer Rate), or an inter bank rate in some other currency environment.

1.1 Elements of Arbitrage Theory and Derivative Pricing

We here review some basic principles of arbitrage theory and derivative pricing. The results are formulated for price systems which are governed by Itô processes, processes which can be represented by stochastic Itô-integrals with respect to Brownian motion. A summary of the involved stochastic calculus and other relevant probabilistic tools is given in Appendix A.1.

1.1.1 Arbitrage-free Systems, Self-Financing Trading Strategies, Complete Markets

Let $(\Omega, \mathcal{F}, P)$ be a probability space with a filtration $\mathbb{F} = (\mathcal{F}_t)_{0 \leq t \leq T_\infty < \infty}$ generated by an $\mathbb{R}^m$-valued Brownian motion W which satisfies the "usual condition". The filtration is thus restricted to a fixed finite interval. Our starting point is an n-dimensional process $B = (B(t))_{0 \leq t \leq T_\infty}$ of tradable securities given by the SDE

$$\frac{dB_i(t)}{B_i(t)} = \mu_i(t)\, dt + \sum_{j=1}^m \sigma_{ij}(t)\, dW_j(t), \qquad B_i(0) > 0, \quad i = 1, \ldots, n. \quad (1.1)$$

For technical reasons, the coefficient processes μ and σ are assumed to be $\mathbb{F}$-predictable (continuous from the left is enough), and such that $\int_0^{T_\infty} |\mu_i(u)|\, du < \infty$ P-a.s. The $\mathbb{R}^m$-valued process $\sigma_i = (\sigma_{ij})_{j=1,\ldots,m}$ is assumed to be W-integrable for each i. This ensures in particular that (1.1) has a unique strong solution with values in $\mathbb{R}^n_{++} := \{x \in \mathbb{R}^n \mid x_i > 0\}$ for $i = 1, \ldots, n$. In our context of modelling interest rate dynamics we think of B as describing the price evolution of n risky bonds traded in a financial market where prices are expressed in some fixed currency unit.

DEFINITION 1.1   *The price system (market) B in (1.1) is said to be* **Arbitrage-free** *(AF) if there exists an adapted process ξ on $(\Omega, \mathcal{F}, P)$ with $\xi > 0$ and $\xi_0 = 1$, such that ξB_i are martingales for all $1 \le i \le n$. The process ξ is called a   state price deflater.*

Note that in (1.1) the *state price deflater* makes deflated prices martingales in the *actual measure*. See also Duffie (2001). From the fact that ξ is adapted to the Brownian filtration $\mathcal{F}$ and the fact that the B_i are positive Itô processes, it is not difficult to show by the martingale representation theorem that ξ satisfies also an Itô SDE,

$$\frac{d\xi(t)}{\xi(t)} = -r(t)\,dt - \sum_{j=1}^{m} \lambda_j(t)\,dW_j(t), \qquad \xi(0) = 1, \qquad (1.2)$$

for a certain predictable scalar process r and vector process λ in $\mathbb{R}^m$. We will refer to r as the (virtual) *short rate* and λ as the *the market price of risk* process.

DEFINITION 1.2   *A* **trading strategy** *in the above market is an $\mathbb{R}^n$-valued $\mathbb{F}$-predictable B-integrable process φ. If φ with corresponding value process*

$$V^\varphi := \sum_{i=1}^{n} \varphi_i B_i =: \varphi^\top B$$

satisfies the self-financing condition

$$V^\varphi(t) = V^\varphi(0) + \int_0^t \varphi(u)^\top dB(u), \qquad 0 \le t \le T_\infty, \qquad (1.3)$$

we speak of a **Self-Financing Trading Strategy** *(SFTS) in the market B.*

It should be noted that in a more general setup one usually requires the processes ξB_i in Definition 1.1 to be local martingales rather than true martingales. For the present exposition, however, we want to avoid technical details related to the issue of local martingales versus true martingales, and therefore we assume in Definition 1.1 more strongly the martingale property. For similar reasons we restrict the class of trading strategies in an arbitrage-free system B to the class of *admissible strategies* in the system B.

DEFINITION 1.3   *For an arbitrage-free system B with price deflater ξ we call a trading strategy φ* **admissible** *if the process $t \to \int_0^t \varphi^\top d(\xi B)$ (which is a local martingale by Theorem A.1) is a true martingale on $[0, T_\infty]$.*

Without going into details, we here just note that a local martingale is a true martingale under certain extra integrability conditions, e.g., Novikov's conditions. The economical relevance of an arbitrage-free price system B is that, within this system, there is no possibility for a "free lunch" in the following sense.

THEOREM 1.1
Assume B in (1.1) is arbitrage-free with price deflater ξ, and $T \leq T_\infty$. If φ is an admissible self-financing trading strategy with initial value $V^\varphi(0) = \varphi(0)^\top B(0) = 0$ and $V^\varphi(T) \geq 0$ a.s., then $V^\varphi(T) = 0$ a.s.

The proof of this theorem is based on the following lemma

LEMMA 1.1
Let φ be a self-financing trading strategy in the system B, i.e., (1.3) holds. Then, for any positive continuous adapted semimartingale ξ such that φ is ξ-integrable we have

$$\varphi(T)^\top \xi(T) B(T) = \varphi(0)^\top \xi(0) B(0) + \int_0^T \varphi(u)^\top d\left(\xi(u) B(u)\right), \qquad (1.4)$$

so φ is also a self-financing trading strategy in the system ξB.

PROOF   By the SFTS property $d(\varphi^\top B) = \varphi^\top dB$ the value process $\varphi^\top B$ is a continuous semimartingale. Then, by applying two times the product rule (A.11) for continuous semimartingales and again the self-financing property, it follows that, in condensed notation,

$$
\begin{aligned}
d\left(\varphi^\top(\xi B)\right) &= \xi d(\varphi^\top B) + \varphi^\top B d\xi + d(\varphi^\top B) \cdot d\xi \\
&= \xi \varphi^\top dB + \varphi^\top B d\xi + (\varphi^\top dB) \cdot d\xi \\
&= \varphi^\top(\xi dB + B d\xi + dB \cdot d\xi) \\
&= \varphi^\top d(\xi B),
\end{aligned}
$$

where $dX \cdot dY := d\langle X, Y \rangle$, with $\langle X, Y \rangle$ being the covariation process of X and Y. □

PROOF   (Theorem 1.1) Since B is arbitrage-free there exists by definition a price deflater $\xi > 0$ such that the ξB_i are martingales, and ξ is continuous by (1.2). By using the admissibility of the SFTS for the system (ξ, B),

Lemma 1.1, and $\varphi(0)^\top B(0) = 0$, it then follows that $E\varphi(T)^\top \xi(T)B(T) = 0$ by taking the expectation of (1.4). Hence, $\varphi(T)^\top \xi(T)B(T) = 0$ a.s. and so $\varphi(T)^\top B(T) = 0$, a.s. since $\xi > 0$. ▯

For an arbitrage-free system the dynamics of asset prices (1.1) and price deflater (1.2) can be written in integrated form,

$$B_i = B_i(0) \exp\left[\int_0^t (\mu_i - \frac{1}{2}|\sigma_i|^2)ds + \int_0^t \sigma_i^\top dW\right],$$

$$\xi = \exp\left[\int_0^t (-r - \frac{1}{2}|\lambda|^2)ds - \int_0^t \lambda^\top dW\right]. \tag{1.5}$$

Equations (1.5) are easily verified by applying Itô's lemma (A.10) to them, which yields (1.1) and (1.2) again. From the representation (1.5) it follows that

$$\xi B_i = B_i(0) \exp\left[\int_0^t (\mu_i - r - \frac{1}{2}|\sigma_i|^2 - \frac{1}{2}|\lambda|^2)ds + \int_0^t (\sigma_i - \lambda)^\top dW\right],$$

and so, since ξB_i are martingales for every i,

$$-\frac{1}{2}|\sigma_i - \lambda|^2 = \mu_i - r - \frac{1}{2}|\sigma_i|^2 - \frac{1}{2}|\lambda|^2,$$

or equivalently,

$$\mu_i = r + \sigma_i^\top \lambda, \quad \text{for} \quad i = 1, .., n. \tag{1.6}$$

We conclude that for an arbitrage-free price system (1.1) the dynamics of a price deflater ξ can be written in the form (1.2), where r and λ satisfy (1.6). Conversely, if there exist for a price system (1.1) predictable processes r and λ such that (1.6) holds, and the solution ξ of (1.2) is such that the processes ξB_i satisfy a certain integrability condition (e.g., Novikov's condition), then the system is arbitrage-free.

We next introduce the concept of completeness which essentially comes down to perfect replicability of all measurable claims.

DEFINITION 1.4   *Let the price system B in (1.1) be arbitrage-free. Then the system is said to be* **complete** *if the price deflater $\xi > 0$, with ξB_i being martingales for all i and $\xi_0 = 1$, is* **unique.**

If we impose a full-rank restriction on the volatility structure of system (1.1), we have the following proposition by Definition 1.4, no-arbitrage condition (1.6), and some linear algebra.

PROPOSITION 1.1
Suppose for each (t, ω) the $n \times m$ matrix σ, defined by $\sigma[i, k] := \sigma_i[k]$ has full rank m, hence $m \leq n$. Then with $\mathbf{1} := [1, \ldots, 1]^\top \in \mathbb{R}^n$ we have

- *For $m = n$, the market is arbitrage-free but incomplete.*

- *If $m = n - 1$ and*

  - **1** $\notin \operatorname{range}(\sigma)$, *the market is arbitrage-free and complete.*
  - **1** $\in \operatorname{range}(\sigma)$,
    * *the market is arbitrage-free but incomplete if $\mu \in span\{\mathbf{1}, \sigma\}$.*
    * *there is arbitrage if $\mu \notin span\{\mathbf{1}, \sigma\}$.*

- *If $m < n - 1$ and*

  - $\mu \in span\{\mathbf{1}, \sigma\}$, *the market is arbitrage-free and*
    * *complete if* **1** $\notin \operatorname{range}(\sigma)$.
    * *incomplete if* **1** $\in \operatorname{range}(\sigma)$.
  - $\mu \notin span\{\mathbf{1}, \sigma\}$, *there is arbitrage.*

The next theorem and Theorem 1.1 are essentially due to Harrison & Pliska (1981,1983) and Delbaen and Schachermayer (1994).

THEOREM 1.2
If the arbitrage-free price system B in (1.1) is complete, then for any T, $0 \leq T \leq T_\infty$, and any random variable $C_T \in \mathcal{F}_T$ (an $\mathcal{F}_T$-claim), there exists an SFTS φ and an initial investment $V^\varphi(0)$ such that

$$C_T = V^\varphi(T) = V^\varphi(0) + \int_0^T \varphi(t)^\top dB(t).$$

The value of $V^\varphi(0)$ is considered as the fair price of the claim.

PROOF We will sketch a proof of Theorem 1.2 under assumption of the full-rank condition in Proposition 1.1. For a fixed but arbitrary claim the martingale $E^{\mathcal{F}_t}(\xi(T)C_T)$ can be represented by the martingale representation theorem as

$$E^{\mathcal{F}_t}(\xi(T)C_T) = E^{\mathcal{F}_0}(\xi(T)C_T) + \int_0^t \xi \alpha^\top dW, \qquad (1.7)$$

for some predictable process α in $\mathbb{R}^m$. For any predictable process φ we have

$$\int_0^t \varphi^\top d(\xi B) = \int_0^t \sum_i \varphi_i \xi B_i (\sigma_i - \lambda)^\top dW. \qquad (1.8)$$

By Proposition 1.1, completeness implies **1** $\notin \operatorname{range}(\sigma)$ and $\operatorname{rank}(\sigma) < n$. So, in particular, the $n \times m$ matrix $\sigma - \mathbf{1}\lambda^\top$ has full rank m also and thus the

$m \times m$ matrix $(\sigma - \mathbf{1}\lambda^\top)^\top(\sigma - \mathbf{1}\lambda^\top)$ is invertible. Therefore, the solution space of the linear system $y^\top(\sigma - \mathbf{1}\lambda^\top) = \alpha^\top$ can be written as

$$y = (\sigma - \mathbf{1}\lambda^\top)\left((\sigma - \mathbf{1}\lambda^\top)^\top(\sigma - \mathbf{1}\lambda^\top)\right)^{-1}\alpha + \mathcal{N},$$

with $\mathcal{N} = \{v : v^\top(\sigma - \mathbf{1}\lambda^\top) = 0\}$. Note that

$$
\begin{aligned}
\mathcal{N}_0 &:= \{v : v^\top(\sigma - \mathbf{1}\lambda^\top) = 0 \ \wedge \ v^\top\mathbf{1} = 0\} \\
&= \{v : v^\top\sigma = 0 \ \wedge \ v^\top\mathbf{1} = 0\} \\
&= \mathrm{span}^\perp\{\mathrm{range}(\sigma), \mathbf{1}\} \subset \mathcal{N},
\end{aligned}
$$

where $0 \le \dim\mathcal{N}_0 = n - m - 1 < \dim\mathcal{N} = n - m$. It thus follows that there exists a predictable process η in $\mathbb{R}^n$ such that $\sum_i \eta_i(\sigma_i - \lambda) = \alpha$ with $\sum_i \eta_i(t) = \eta(t)^\top\mathbf{1} = \xi^{-1}(t)E^{\mathcal{F}_t}\left(\xi(T)C_T\right)$, namely,

$$\eta = (\sigma - \mathbf{1}\lambda^\top)\left((\sigma - \mathbf{1}\lambda^\top)^\top(\sigma - \mathbf{1}\lambda^\top)\right)^{-1}\alpha + v,$$

with

$$v^\top\mathbf{1} = \xi^{-1}(t)E^{\mathcal{F}_t}\left(\xi(T)C_T\right) - \mathbf{1}^\top(\sigma - \mathbf{1}\lambda^\top)\left((\sigma - \mathbf{1}\lambda^\top)^\top(\sigma - \mathbf{1}\lambda^\top)\right)^{-1}\alpha.$$

Moreover, if $m = n - 1$ the solution process η is unique.

Let us now consider the trading strategy $\varphi_i = \eta_i/B_i$ for the system B. We will show that φ is self-financing and that $\varphi^\top(T)B(T) = C_T$, hence the corresponding self-financing portfolio replicates the claim. From (1.7) and (1.8) we obtain

$$
\begin{aligned}
\int_0^t \varphi^\top d(\xi B) &= \int_0^t \xi\alpha^\top dW \\
&= E^{\mathcal{F}_t}\left(\xi(T)C_T\right) - E^{\mathcal{F}_0}\left(\xi(T)C_T\right) \\
&= \xi(t)\varphi^\top(t)B(t) - \xi(0)\varphi^\top(0)B(0);
\end{aligned}
$$

hence the pair $(\varphi, \xi B)$ satisfies the self-financing condition (1.3). Then, applying Lemma 1.1 with ξ replaced by ξ^{-1} and B replaced by ξB yields that the pair (φ, B) satisfies the self-financing condition (1.3) as well. So (φ, B) is a self-financing portfolio and

$$
\begin{aligned}
\varphi(t)^\top B(t) &= \xi^{-1}(t)E^{\mathcal{F}_t}\left(\xi(T)C_T\right) \\
&= \xi^{-1}(0)E^{\mathcal{F}_0}\left(\xi(T)C_T\right) + \int_0^t \varphi^\top dB,
\end{aligned}
$$

where in particular $\varphi(T)^\top B(T) = C_T$ and $\xi^{-1}(0)E^{\mathcal{F}_0}\left(\xi(T)C_T\right)$ is considered to be the fair price of the claim at time $t = 0$. □

REMARK 1.1   In the case where the system (1.1) constitutes a complete market (see Proposition 1.1), it can be shown along the same lines as in the proof of Theorem 1.2 (e.g., see also Reiß, Schoenmakers, Schweizer 2001) that there exists a portfolio β_t generated by some self-financing trading strategy, such that $\beta(t) = \beta(0) \exp(\int_0^t r(s)ds)$; hence β plays the role of a savings account . However, following Jamshidian (1997), we explain in the next section that assuming the existence of such a savings account is not necessary for pricing and hedging of Libor derivative products. $\square$

1.1.2 Derivative Claim Pricing in Different Measures

As we saw in Section 1.1.1 (Theorem 1.2), in a complete market any measurable claim C_T can be replicated by a self-financing portfolio and the initial value of this portfolio is the fair price of the claim. If the market is arbitrage-free but incomplete, the collection of replicable claims is generally smaller but not empty. For instance, any claim of type $C_T = \sum_i^n \alpha_i B_i(T)$ with constant α's is trivially replicable.

Let us suppose C_T is a replicable claim in an arbitrage-free system (1.1) and let φ be a self-financing trading strategy which replicates the claim, i.e.,

$$C_T = \varphi(T)^\top B(T) = \varphi(0)^\top B(0) + \int_0^T \varphi(t)^\top dB(t).$$

Since (1.1) is arbitrage-free there exists a price deflater ξ such that ξB_i are martingales. By Lemma 1.1 we have

$$\xi(T)C_T = \xi(T)\varphi(T)^\top B(T) = \xi(0)\varphi(0)^\top B(0) + \int_0^T \varphi(s)^\top d\left(\xi(s)B(s)\right),$$

and by the martingale property of the process $\int_0^t \varphi^\top d\left(\xi B\right)$ we get for $t \leq T$,

$$E^{\mathcal{F}_t}\left(\xi(T)C_T\right) = \xi(t)\varphi(t)^\top B(t) = \xi(0)\varphi(0)^\top B(0) + \int_0^t \varphi(s)^\top d\left(\xi(s)B(s)\right),$$

hence,

$$\varphi(t)^\top B(t) = \frac{1}{\xi(t)} E^{\mathcal{F}_t}\left(\xi(T)C_T\right). \qquad (1.9)$$

From (1.9) we conclude the following. On the one hand, in an incomplete market where the price deflater process ξ is not unique, the right-hand side of (1.9) does not depend on ξ. On the other hand, if $\widetilde{\varphi}$ is another SFTS which replicates C_T, we have $\varphi(t)^\top B(t) = \widetilde{\varphi}(t)^\top B(t)$, i.e., the two SFTS's have always the same price. Thus, the time $t < T$ value of the claim C_T is properly defined by

$$C_t := \frac{1}{\xi(t)} E^{\mathcal{F}_t}\left(\xi(T)C_T\right). \qquad (1.10)$$

As a result, in a complete market where any $\mathcal{F}_T$-measurable claim C_T can be hedged by an SFTS (Theorem 1.2), the price C_t of this claim at a prior time $t < T$ is given by (1.10).

In (1.10) the price of a derivative claim is represented with respect to the objective or "real world" measure. We now introduce the notion of *numeraire measures* and will give representations for the claim price C_t in different numeraire measures by using numeraire transformations.

DEFINITION 1.5   *Let the system (1.1) be arbitrage-free and ξ be a price deflater such that ξB_i are martingales. Let further $A > 0$ be a positive adapted process such that ξA is a martingale. The A-**numeraire measure** P_A is then defined via the Radon-Nykodym derivative*

$$\frac{dP_A}{dP} := \frac{\xi(T_\infty)A(T_\infty)}{A(0)}, \qquad by$$

$$E_A X := \int X dP_A := \int X \frac{dP_A}{dP} dP = E\left(X \frac{\xi(T_\infty)A(T_\infty)}{A(0)}\right),$$

for any $\mathcal{F}$-measurable random variable X. Moreover, if X_t is $\mathcal{F}_t$-measurable, we have by the martingale property of ξA,

$$E_A X_t = E E^{\mathcal{F}_t}\left(X_t \frac{\xi(T_\infty)A(T_\infty)}{A(0)}\right) = E\left(X_t \frac{\xi(t)A(t)}{A(0)}\right) = E\left(X_t M_A(t)\right),$$

where the Radon-Nykodym process

$$M_A(t) := E^{\mathcal{F}_t} \frac{dP_A}{dP} = \frac{\xi(t)A(t)}{A(0)}$$

is a martingale associated with the numeraire A.

The following lemma is quite useful and easy to prove.

LEMMA 1.2
If ξA and ξX are martingales then X/A is a P_A martingale.

PROOF   By a well-known rule for the conditional expectation of a Radon-Nykodym transformed measure, we have for $0 \le t \le t + s \le T_\infty$,

$$E_A^{\mathcal{F}_t} \frac{X(t+s)}{A(t+s)} = \frac{E^{\mathcal{F}_t}\left(\frac{dP_A}{dP} \frac{X(t+s)}{A(t+s)}\right)}{E^{\mathcal{F}_t} \frac{dP_A}{dP}} = \frac{E^{\mathcal{F}_t}\left(M_A(t+s)X(t+s)/A(t+s)\right)}{M_A(t)}$$

$$= \frac{E^{\mathcal{F}_t}\left(\xi(t+s)X(t+s)/A(0)\right)}{\xi(t)A(t)/A(0)} = \frac{X(t)}{A(t)}.$$

$\square$

We now have a representation for the option price (1.10) in terms of the numeraires A in Definition 1.5, due to the following proposition.

PROPOSITION 1.2

As in Definition 1.5, let $A > 0$ and ξA be a martingale. Let further C_T be an option ($\mathcal{F}_T$-claim) which can be hedged by an SFTS. Then, we have

$$C_t = \frac{1}{\xi(t)} E^{\mathcal{F}_t}\left(\xi(T)C_T\right)$$

$$= A(t) E_A^{\mathcal{F}_t} \frac{C_T}{A(T)}. \qquad (1.11)$$

PROOF Since ξC and ξA are martingales it follows from lemma (1.2) that C/A is a P_A martingale, hence $C_t/A(t) = E_A^{\mathcal{F}_t}\left(C_T/A(T)\right)$. ⬚

Proposition 1.2 is particularly useful in the case where A is one of the assets in the system (1.1), i.e., $A = B_i$ for some i. Moreover, if B_i is a T_i-maturity zero coupon bond with $B_i(T) = 1$, we get simply $C_t = B_i(t) E_{B_i}^{\mathcal{F}_t} C_T$.

1.2 Modelling of Effective Forward Rates

In this section we will derive the dynamics of forward Libor and swap rate processes defined with respect to a fixed system of tenor dates. Since both Libor and swap rates can be expressed in terms of zero-coupon bonds, it is natural to derive their respective dynamics from an arbitrage-free system (1.1) consisting of (zero) bonds B_i. The Libor rate process will be studied in Section 1.2.1 and the swap rate dynamics in Section 1.2.2.

1.2.1 Libor Rate Processes and Measures

Let us consider a fixed sequence of tenor dates $0 < T_1 < T_2 < \cdots < T_n < T_\infty$, called a tenor structure, together with a sequence of so called day-count fractions $\delta_i := T_{i+1} - T_i$, $i = 1, \ldots, n - 1$. With respect to this tenor structure we consider an arbitrage-free system (1.1) of zero-bond processes B_i, for $i = 1, \ldots, n$, where each B_i lives on the interval $[0, T_i]$ and ends up with its face value $B_i(T_i) = 1$. With respect to this bond system we deduce a system of forward rates, called Libor rates, which are defined by

$$L_i(t) := \frac{1}{\delta_i}\left(\frac{B_i(t)}{B_{i+1}(t)} - 1\right), \qquad 0 \le t \le T_i, \quad 1 \le i \le n - 1. \qquad (1.12)$$

Note that L_i in (1.12) is the annualized effective forward rate to be contracted for at the (present) date t, for a loan over a forward period $[T_i, T_{i+1}]$. Based on this rate one has to pay at T_{i+1} an interest amount of $\$\delta_i L_i(T_i)$ on a $\$1$ notional.

We will now derive the dynamics of the Libor process L when the bond dynamics is given by the system (1.1) under the no-arbitrage condition (1.6). By applying Itô via Lemma A.2 to the Libor definition (1.12), using (1.6) and (1.12) again, we obtain straightforwardly

$$dL_i = \delta_i^{-1}(1 + \delta_i L_i)\left((\mu_i - \mu_{i+1} - \sigma_{i+1}^{\top}(\sigma_i - \sigma_{i+1}))\,dt + (\sigma_i - \sigma_{i+1})^{\top}dW\right)$$
$$= \delta_i^{-1}(1 + \delta_i L_i)(\sigma_i - \sigma_{i+1})^{\top}\left(dW + (\lambda - \sigma_{i+1})dt\right), \qquad 1 \le i < n.$$

By introduction of the Libor volatility processes $\gamma_i \in \mathbb{R}^m$,

$$L_i \gamma_i := \delta_i^{-1}(1 + \delta_i L_i)(\sigma_i - \sigma_{i+1}), \qquad (1.13)$$

and the drifted Brownian motions

$$dW^{(j)} := dW + (\lambda - \sigma_j)dt, \qquad 1 \le j \le n, \qquad (1.14)$$

we may write for $1 \le i < n$,

$$dL_i = L_i \gamma_i^{\top} dW^{(i+1)} \qquad (1.15)$$
$$= -\sum_{j=i+1}^{n-1} \frac{\delta_j L_i L_j}{1 + \delta_j L_j} \gamma_i^{\top} \gamma_j dt + L_i \gamma_i^{\top} dW^{(n)}, \quad 0 \le t \le T_i. \qquad (1.16)$$

From Definition 1.12 and Lemma 1.2 it follows that L_i is a martingale with respect to the measure $P_{B_{i+1}}$. From a general martingale representation theorem it then follows that $\left(W^{(i+1)}(t)\right)_{0 \le t \le T_i}$ in (1.15) is standard Brownian motion under $P_{B_{i+1}}$, for $1 \le i < n$. Similarly, since again by Lemma 1.2 $B_{i+1}/B_i = (1 + \delta_i L_i)^{-1}$ is a martingale under P_{B_i}, we can show that $\left(W^{(i)}(t)\right)_{0 \le t \le T_i}$ is standard Brownian motion under P_{B_i}, for $1 \le i < n$. By combining, we obtain the following corollary.

COROLLARY 1.1
For $j = 1, \ldots, n$ the process

$$W^{(j)}(t) = W(t) + \int_0^t (\lambda - \sigma_j)ds, \quad 0 \le t \le T_j \wedge T_{n-1},$$

is an m-dimensional Brownian motion under the measures P_{B_j}.

REMARK 1.2   For each particular i the zero bond B_i terminates at T_i and so, after passing a maturity T_i, the number of "living bonds" in the bond system (1.1) decreases with one. As a consequence, the arbitrage theory in Section 1.1 should be considered on the time interval $[0, T_1]$ with respect to the full system $B_1, \ldots, B_n$, then on the time interval $(T_1, T_2]$ with respect to $B_2, \ldots, B_n$, and so on. In order to meet the full-rank condition in Proposition 1.1 in each such interval, it is natural to assume at most n Brownian motions, hence $m \le n$, and $\sigma[i, k] = \sigma_i[k] = 0$, for $i - k > n - m$.   □

From now on we will use for the numeraire measure P_{B_j} the shorter notation P_j and for the related expectation we will write E_j. The Libor dynamics (1.16) is derived from a given arbitrage-free zero bond system where the volatility process γ can be expressed explicitly in the bond process via (1.13) and (1.12). Now the following question arises. Let us take some Libor volatility process γ. Does there exist an arbitrage-free zero-bond process (1.1) such that (1.12) holds? The following proposition gives a partially confirmative answer sufficient for our purposes (see Jamshidian (1997), Th. 5.3 and 7.1, for more details).

PROPOSITION 1.3

Existence of the Libor process *Suppose we are given a bounded and locally Lipschitz volatility structure of type $\gamma_i(t, L)$, $1 \le i < n$. Then there exists an arbitrage-free system B of bond prices (1.1) satisfying $B_i(T_i) = 1$ and a price deflater ξ (i.e., ξB_i are P-martingales for $1 \le i \le n$), such that (1.16) with $\gamma_i = \gamma_i(t, L)$ has a unique solution for which (1.12) holds.*

Let us now start with a given bounded and locally Lipschitz volatility function $\mathbb{R} \times \mathbb{R}^{n-1} \ni (t, L) \to \gamma(t, L) \in \mathbb{R}^{(n-1) \times m}$. Consider the corresponding Libor process $L(t)$ as a solution of SDE (1.16), where $W^{(n)}$ is an m-dimensional Brownian motion with respect to P_n, starting at some $L(0) \in \mathbb{R}^{n-1}_+$. In fact, economically only the distribution of the Libor process is relevant. For this reason we assume without loss of generality that we have at most $n-1$ Brownian motions in (1.16), i.e., $m \le n-1$, and that $\gamma(t, L)$ has full rank m. Indeed, if $m \ge n$ it would be possible to generate by (1.16) a Libor process with the same distribution, using another full rank volatility function with $m \le n - 1$. Due to Proposition 1.3, there exist at the background an arbitrage-free zero bond system (1.1) which satisfies $B_i(T_i) = 1$, $1 \le i \le n$, and (1.12). As a consequence, all ratios B_i/B_j are measurable with respect to L, hence the filtration generated by the m (at most $n - 1$) Brownian motions. Moreover, in this setup the following basic result, essentially due to Jamshidian (1997) Th. 5.2, applies.

THEOREM 1.3

Hedging Libor derivative claims *Consider an arbitrage-free bond system (1.1) and the corresponding Libor process (1.12) with dynamics given by (1.16). Suppose that $W^{(n)}$ is an m-dimensional Brownian motion with $m < n$ and the volatility process γ has constant rank m and is predictable with respect to $\mathcal{J}$, the filtration generated by $W^{(n)}$. Then, any claim C_T such that $C_T/B_n(T)$ is measurable with respect to $\mathcal{J}_T$ can be priced and hedged by a self-financing trading strategy φ in the bond system B. The arbitrage-free price of* *the claim is given by (see also (1.11))*

$$C_t := \varphi(t)^\top B(t) = B_n(t) E_n^{\mathcal{J}_t} \frac{C_T}{B_n(T)}. \qquad (1.17)$$

PROOF   We will sketch the proof which is similar to the proof of Theorem 1.2. By the martingale representation theorem there exists a $\mathcal{J}$-predictable process $\beta \in \mathbb{R}^m$ such that

$$\frac{C_t}{B_n(t)} = E_n^{\mathcal{J}_t} \frac{C_T}{B_n(T)} = E^{\mathcal{J}_0} \frac{C_T}{B_n(T)} + \int_0^t \beta^\top dW^{(n)}.$$

Since L is $\mathcal{J}$-adapted, the system $(B/B_n) := (B_i/B_n)_{1 \le i < n} \in \mathbb{R}^{n-1}$ is a $\mathcal{J}$-adapted martingale in P_n, and so $d(B/B_n) = \nu dW^{(n)}$ for some predictable process $\nu \in \mathbb{R}^{(n-1) \times m}$. It is not difficult to see that since γ has full rank, ν has full rank also, and so we may consider $\theta := \nu(\nu^\top \nu)^{-1}\beta \in \mathbb{R}^{n-1}$ yielding

$$\theta^\top d(B/B_n) = \beta^\top (\nu^\top \nu)^{-1} \nu^\top \nu dW^{(n)} = \beta^\top dW^{(n)} = d(C/B_n).$$

By taking $\varphi_i := \theta_i$, for $1 \le i < n$ and $\varphi_n := C/B_n - \theta^\top(B/B_n)$, we thus have

$$d\left(((\varphi^\top B)/B_n\right) = d(C/B_n) = \theta^\top d(B/B_n) = \sum_{i=1}^n \varphi_i d(B_i/B_n).$$

Hence, by Lemma 1.1 it follows that the pair (φ, B) constitutes a self-financing portfolio with

$$C_t = B_n(t) E_n^{\mathcal{J}_t} \frac{C_T}{B_n(T)} = \varphi(t)^\top B(t) = \varphi(0)^\top B(0) + \int_0^t \varphi^\top dB.$$

$\square$

Note that in Theorem 1.3 completeness of the Bond system (1.1) is not required. As we will see, Theorem 1.3 applies to almost all Libor derivative claims in practice.

From (1.13) and Corollary 1.1 we easily derive for any i, $1 \le k \le n$,

$$dW^{(n)} = dW^{(k)} + \sum_{j=k}^{n-1} \frac{\delta_j L_j \gamma_j}{1 + \delta_j L_j} dt, \quad 0 \le t \le T_i \wedge T_{n-1}, \qquad (1.18)$$

where an empty sum is defined to be 0. Thus, by inserting (1.18) in (1.16) we obtain the dynamics of L_i in any bond measure P_k, $1 \le k \le n$,

$$dL_i = - \sum_{j=i+1}^{n-1} \frac{\delta_j L_i L_j}{1 + \delta_j L_j} \gamma_i^\top \gamma_j dt + \sum_{j=k}^{n-1} \frac{\delta_j L_i L_j}{1 + \delta_j L_j} \gamma_i^\top \gamma_j dt +$$

$$+ L_i \gamma_i^\top dW^{(k)}, \quad 0 \le t \le T_i \wedge T_k \wedge T_{n-1}, \qquad (1.19)$$

where $W^{(k)}$ is a Brownian motion under P_k. Note that for $k > i+1$, the drift term in (1.19) is negative, for $k = i+1$ equal to zero (see (1.15)), and for $k \leq i$ positive. In practical situations it makes sense to choose the bond measure P_k such that the drift term in (1.19) is as small as possible. For instance, when a claim involves only L_i, it is natural to take as pricing measure P_{i+1}. Usually, however, one needs a Libor measure which lives on the whole tenor structure like the forward bond or terminal measure P_n. But, the drift of L_i in this measure contains $n - i$ terms and so may be large when i is small. This may cause undesirable numerical errors (e.g., in an SDE simulation with time discretization steps $\Delta t_k = \delta_k$). Therefore, as an alternative, we introduce a new measure on the whole tenor structure, called the *spot Libor measure*.

The spot Libor measure

As mentioned above, we do not explicitly assume a continuous savings account

$$\beta_t := e^{\int_0^t r(s)ds},$$

hence a tradable short rate r, since this is not necessary for pricing Libor derivative claims. As an alternative we consider a discrete savings account generated by the spot Libor rate in the following way. We start at $t = T_0 = 0$ with an initial capital \$1 invested in the zero bond B_1. At T_1 we reinvest (roll over) the account value $1/B_1(0)$ in zero bond B_2, yielding an account value

$$\frac{1}{B_1(0)B_2(T_1)} = \frac{1}{B_1(0)}(1 + \delta_1 L_1(T_1)),$$

at T_2 and so on. We thus end up with the so called spot Libor rolling over account,

$$B_*(0) := 1, \qquad B^*(t) := \frac{B_{m(t)}(t)}{B_1(0)} \prod_{i=1}^{m(t)-1} (1 + \delta_i L_i(T_i)), \qquad 0 < t \leq T_n,$$

(1.20)

where $m(t) := \min\{m : T_m \geq t\}$ denotes the next reset date at time t. Clearly, the process B_* is a continuous and completely determined by the Libors at the tenor dates. Moreover, B_* can be seen as the value process of a self financing portfolio (φ, B), $B^* = \varphi^\top B$, where

$$\varphi_i(t) = \frac{1_{(T_{i-1}, T_i]}(t)}{B_1(0)} \prod_{i=1}^{m(t)-1} (1 + \delta_i L_i(T_i)), \qquad 1 < i \leq n, \ 0 < t \leq T_n,$$

with $\varphi_i(0) := \varphi_i(0+)$. Hence, φ is left continuous and by Lemma 1.1 it follows that ξB_* is a martingale. So we may introduce the *spot Libor measure P^** according to Definition 1.5 by

$$\frac{dP_*}{dP} := \frac{\xi(T_n)B^*(T_n)}{B_*(0)} = \frac{\xi(T_n)}{B_1(0)} \prod_{i=1}^{n-1} (1 + \delta_i L_i(T_i)).$$

Note that B^* is only defined on $[0, T_n]$, so here we have to replace T_∞ in Definition 1.5 by T_n.

Let us now derive the Libor dynamics in P_*. From (1.20) it follows that

$$\frac{dB_*(t)}{B_*(t)} 1_{(T_{i-1}, T_i]}(t) = 1_{(T_{i-1}, T_i]}(t) \frac{dB_i(t)}{B_i(t)}$$

$$= 1_{(T_{i-1}, T_i]}(t)(\mu_i dt + \sigma_i^\top dW), \quad 1 < i \le n$$

(in the objective measure), hence the SDE

$$\frac{dB_*(t)}{B_*(t)} = \mu_{m(t)} dt + \sigma_{m(t)}^\top dW, \quad 0 \le t \le T_n,$$

describes the dynamics of the spot Libor rolling-over account B_*. By straight-forward application of Itô via Lemma A.2 and using (1.6) we have for any j,

$$d\frac{B_j(t)}{B_*(t)} = \frac{B_j(t)}{B_*(t)} \left((\mu_j - \mu_{m(t)} - (\sigma_j - \sigma_{m(t)})^\top \sigma_{m(t)}) dt + (\sigma_j - \sigma_{m(t)})^\top dW \right)$$

$$= \frac{B_j(t)}{B_*(t)} (\sigma_j - \sigma_{m(t)})^\top \left((\lambda - \sigma_{m(t)}) dt + dW \right)$$

$$=: \frac{B_j(t)}{B_*(t)} (\sigma_j - \sigma_{m(t)})^\top dW^*, \quad 0 \le t \le T_j.$$

Due to Lemma 1.2, B_i/B_* is a P_*-martingale and so W^* is standard Brownian in P_*. Then, by (1.14) we may write

$$dW^{(n)} = dW^* + (\sigma_{m(t)} - \sigma_n) dt, \quad (1.21)$$

and by plugging (1.21) into (1.16) and using (1.13) we obtain the Libor dynamics in P_*,

$$dL_i = \sum_{j=m(t)}^{i} \frac{\delta_j L_i L_j}{1 + \delta_j L_j} \gamma_i^\top \gamma_j \, dt + L_i \gamma_i^\top dW^*, \quad 0 \le t \le T_i. \quad (1.22)$$

We see that in the spot Libor measure L_i in (1.22) contains $i - m(t)$ drift terms, whereas the SDE in the terminal measure (1.16) contains $n - i$ drift terms. For numerical reasons it is important to keep the Libor drift as small as possible. Therefore, loosely speaking, for products involving only short maturity Libors the dynamics in the spot Libor measure (1.22) is preferable whereas for longer dated products the representation in the terminal measure (1.16) may be recommended.

For practical applications a very important class of Libor models consists of the so called Libor *Market* models:

DEFINITION 1.6   *A Libor model where the volatility process $t \to \gamma(t)$ is a deterministic function is called a* **Libor Market Model (LMM).**

From (1.15) we see that in a Libor market model L_i is a *log-normal martingale* in the measure P_{i+1} for $1 \leq i < n$. This is one of the main features of a Libor market model because, as a consequence, important liquidly traded market instruments such as caps (see Section 1.3.1) can be priced with closed form (Black-Scholes type) expressions.

1.2.2 Swap Rate Processes and Measures

An interest rate payer swap on a loan over a period $[T_p, T_q]$, with notional amount $1, is a contract to pay a fixed rate κ and to receive spot Libor (floating) at the settlement dates $T_{p+1}, \ldots, T_q$. By a simple portfolio argument we can show that the arbitrage-free value of this contract at $t \leq T_p$ is equal to

$$Swap_{p,q;\kappa}(t) := B_p(t) - B_q(t) - \sum_{j=p}^{q-1} \kappa \delta_j B_{j+1}. \tag{1.23}$$

Indeed, on the one hand, the fixed interest rate payments $ $\kappa \delta_j$ at the dates $T_{p+1}, \ldots, T_q$ have at time t the value of $\sum_{j=p}^{q-1} \kappa \delta_j B_{j+1}$. On the other hand, the future floating payments to be received are at time t equivalent to a portfolio consisting of a zero bond B_p *long* and B_q *short*, hence $B_p - B_q$. As a result, the difference (1.23) is the value of the swap contract at time t.

The *swap rate* $S_{p,q}$ is defined as *that* fixed rate κ for which in (1.23) $Swap_{p,q;\kappa} = 0$. Hence,

$$S_{p,q}(t) = \frac{B_p(t) - B_q(t)}{\sum_{j=p}^{q-1} \delta_j B_{j+1}(t)} =: \frac{B_p(t) - B_q(t)}{B_{p,q}(t)}, \tag{1.24}$$

where $B_{p,q}$ is the so called annuity numeraire.

We refer to (1.24) as the standard swap rate. In practice, however, a swap contract may be defined such that fixed rates need to be settled only at a real subset of the Libor tenor dates in the respective swap period. For example, in the US, UK, and Japanese market Libors are quarterly, whereas swaps are annually settled. In the Euro market Libors are semi-annual, whereas swaps are annually settled. In order to cover these situations we consider a generalized swap contract over a period $[T_p, T_q]$, for a fixed choice of p and q, where the settlement dates are given by a sequence $(T_{p_1}, \ldots, T_{p_k})$ with $T_p =: T_{p_0} < T_{p_1} < \cdots < T_{p_k} = T_q$. This leads to the corresponding generalized swap contract

$$Swap_{p,q;\kappa}^{\mathrm{gen}}(t) := B_p(t) - B_q(t) - \sum_{j=0}^{k-1} \kappa(T_{p_{j+1}} - T_{p_j}) B_{p_{j+1}}(t), \tag{1.25}$$

and generalized swap rate

$$S_{p,q}^{\mathrm{gen}}(t) = \frac{B_p(t) - B_q(t)}{\sum_{j=0}^{k-1}(T_{p_{j+1}} - T_{p_j}) B_{p_{j+1}}(t)} =: \frac{B_p(t) - B_q(t)}{B_{p,q}^{\mathrm{gen}}(t)}, \tag{1.26}$$

with $B_{p,q}^{\text{gen}}$ being the generalized annuity numeraire. For example, taking $p_j = p + j$, $j = 0, \ldots, k := q - p$, yields the standard swap rate again. Henceforth we drop the superscript (gen) since it will be always clear from the context which kind of swap rate is under consideration.

Let us consider the dynamics of the (generalized) annuity numeraire in (1.26). We may write

$$
\begin{aligned}
dB_{p,q} &= \sum_{j=0}^{k-1} (T_{p_{j+1}} - T_{p_j}) dB_{p_{j+1}} \\
&= B_{p,q} \sum_{j=0}^{k-1} \frac{(T_{p_{j+1}} - T_{p_j}) B_{p_{j+1}}}{B_{p,q}} (\mu_{p_{j+1}} dt + \sigma_{p_{j+1}}^\top dW) \\
&= B_{p,q} \left(\sum_{j=0}^{k-1} w_j^{p,q} \mu_{p_{j+1}} dt + \sum_{j=0}^{k-1} w_j^{p,q} \sigma_{p_{j+1}}^\top dW \right),
\end{aligned}
$$

with $w_j^{p,q} = \dfrac{(T_{p_{j+1}} - T_{p_j}) B_{p_{j+1}}}{B_{p,q}}$. By using Itô via Lemma A.2 we then obtain for any $p \le r \le q$,

$$
\begin{aligned}
\frac{d(B_r / B_{p,q})}{(B_r / B_{p,q})} &= \left(\mu_r - \sum_{j=0}^{k-1} w_j^{p,q} \mu_{p_{j+1}} - \sum_{l=0}^{k-1} w_l^{p,q} \sigma_{p_{l+1}}^\top \left(\sigma_r - \sum_{j=0}^{k-1} w_j^{p,q} \sigma_{p_{j+1}} \right) \right) dt \\
&\quad + \left(\sigma_r - \sum_{j=0}^{k-1} w_j^{p,q} \sigma_{p_{j+1}} \right)^\top dW.
\end{aligned}
$$

Since the $w_j^{p,q}$ are weights, i.e., $\sum_{j=0}^{k-1} w_j^{p,q} = 1$, we get by using (1.6),

$$
\begin{aligned}
\frac{d(B_r / B_{p,q})}{(B_r / B_{p,q})} &= \sum_{j=0}^{k-1} w_j^{p,q} (\sigma_r - \sigma_{p_{j+1}})^\top \left(\lambda dt - \sum_{l=0}^{k-1} w_l^{p,q} \sigma_{p_{l+1}} dt + dW \right) \\
&=: \sum_{j=0}^{k-1} w_j^{p,q} (\sigma_r - \sigma_{p_{j+1}})^\top dW^{p,q}. \quad (1.27)
\end{aligned}
$$

Both ξB_r and $\xi B_{p,q}$ are martingales, so $W^{p,q}$ is standard Brownian motion in the measure $P_{p,q}$, induced by the annuity numeraire $B_{p,q}$ according to Lemma 1.2. By now applying (1.27) for $r = p$ and $r = q$ to the swap rate defining expression (1.26), respectively, we obtain

$$
dS_{p,q} = \left(\frac{B_p}{B_{p,q}} \sum_{j=0}^{k-1} w_j^{p,q} (\sigma_p - \sigma_{p_{j+1}}) - \frac{B_q}{B_{p,q}} \sum_{j=0}^{k-1} w_j^{p,q} (\sigma_q - \sigma_{p_{j+1}}) \right)^\top dW^{p,q}
$$

$$
= S_{p,q} \left(\sum_{j=0}^{k-1} w_j^{p,q} (\sigma_p - \sigma_{p_{j+1}}) + \frac{B_q}{B_p - B_q} (\sigma_p - \sigma_q) \right)^\top dW^{p,q}
$$

$$
=: S_{p,q} \sigma_{p,q}^\top dW^{p,q}, \tag{1.28}
$$

where $\sigma_{p,q}$ is the relative volatility process of the swap rate $S_{p,q}$.

Analogue to the definition of a Libor market model we now introduce the notion of a swap market model.

DEFINITION 1.7 *If in (1.28) the volatility process $t \to \sigma_{p,q}(t)$ is a deterministic function we speak of a* **Swap Market Model** *for $S_{p,q}$.*

In a swap market model $\sigma_{p,q}$ is a *log-normal martingale* in the measure $P_{p,q}$. As an important consequence, European options on swap contracts over $[T_p, T_q]$, called *swaptions*, can be priced exactly with Black-Scholes type formulas; see Section 1.3.2. Moreover, as we will show, there exist very accurate swaption approximation formulas for swaptions in a Libor market model (see Section 1.3.3).

By using (1.13) we may express $\sigma_{p,q}$ in (1.28) in the Libor volatility structure to yield

$$
\sigma_{p,q} = \sum_{j=0}^{k-1} w_j^{p,q} \sum_{l=p}^{p_{j+1}-1} \frac{\delta_l L_l}{1 + \delta_l L_l} \gamma_l + \frac{B_q}{B_p - B_q} \sum_{l=p}^{q-1} \frac{\delta_l L_l}{1 + \delta_l L_l} \gamma_l
$$

$$
= \sum_{l=p}^{q-1} \gamma_l \frac{\delta_l L_l}{1 + \delta_l L_l} \left(\sum_{j=j_l}^{k-1} w_j^{p,q} + \frac{B_q}{B_p - B_q} \right) \tag{1.29}
$$

with $j_l := \min\{ j : p_{j+1} \geq l+1 \}$. Expression (1.29) will be the starting point for deriving swaption approximation formulas in a Libor market model in Section 1.3.3.

Libor and swap market models are in fact inconsistent with each other. For example, if L_1 and L_2 have deterministic volatilities it follows immediately from (1.29) that the volatility of the standard swap $S_{1,3}$ is non-deterministic. Therefore one usually chooses that market model which is most appropriate for pricing and hedging a particular product under consideration.

REMARK 1.3 In a Libor market model where $n-1$ Libor volatilities γ_i are specified by deterministic vector functions, the bond volatility differences

$\sigma_i - \sigma_k$ are implicitly determined via (1.13). Hence the bond volatilities are determined up to one degree of freedom. In contrast, for swap market models in general one cannot specify simultaneously $n(n-1)/2$ deterministic swaption volatilities $\sigma_{p,q}$ for *all* $1 \le p < q \le n$, since regarding (1.28) we have only n bond volatilities to determine them, hence n degrees of freedom. But, for instance, for a complete system of standard swaps it is possible to choose $\sigma_{1,n}, \ldots, \sigma_{n-1,n}$ simultaneously deterministic. ▯

1.3   Pricing of Caps and Swaptions in Libor and Swap Market Models

1.3.1   Libor Caps and Caplets

A Libor cap on a loan over period $[T_p, T_q]$, with notional amount \$1, is a contract to pay at the settlement dates $T_{p+1}, \ldots, T_q$, floating spot Libor capped with a pre-specified strike rate κ. Clearly, this contract is equivalent to a sequence of cash-flows $(L_j(T_j) - \kappa)^+\delta_j$, $p \le j < q$, at the respective settlement dates. Every T_{j+1} pay-off is homogeneous of degree 1, since it can be written as $B_{j+1}(T_{j+1})(L_j(T_j) - \kappa)^+\delta_j$. So, under the measurability conditions in Theorem 1.3 (which are trivially satisfied for Libor volatility structures of type $\gamma(t, L)$), the value of this contract at $t \le T_p$ is given by

$$Cap_{p,q;\kappa}(t) := \sum_{j=p}^{q-1} B_{j+1}(t)E_{j+1}^{\mathcal{J}_t}\left(L_j(T_j) - \kappa\right)^+\delta_j, \qquad (1.30)$$

where $(\mathcal{J}_t)_{0 \le t \le T_{n-1}}$ is the filtration generated by the Libor process. In (1.30) each T_{j+1} cash-flow can be seen as the cash-flow of a "mini cap" over T_j, T_{j+1}, called a *caplet*, and is priced in its canonical measure P_{j+1} via Proposition 1.2. In a Libor market model caplets can be priced with Black-Scholes type expressions and so, by (1.30), caps also.

PROPOSITION 1.4
The price of a caplet over $[T_j, T_{j+1}]$ with strike κ in a LMM with deterministic volatility structure $t \to \gamma(t)$ is given by

$$\begin{aligned} Caplet_{j,j+1;\kappa}(t) &= B_{j+1}(t)E_{j+1}^{\mathcal{J}_t}\left(L_j(T_j) - \kappa\right)^+\delta_j \qquad (1.31) \\ &= B_{j+1}(t)(\delta_j L_j(t)\mathcal{N}(d_+) - \delta_j \kappa \mathcal{N}(d_-)), \end{aligned}$$

with $\mathcal{N}$ being the standard normal distribution and

$$d_\pm = \frac{\ln(L_j(t)/\kappa) \pm (\sigma_j^B)^2(T_j - t)/2}{\sigma_j^B \sqrt{T_j - t}}, \qquad (\sigma_j^B)^2 = \frac{1}{T_j - t}\int_t^{T_j} |\gamma_j|^2(s)ds.$$

PROOF   Since γ_i is deterministic, L_j is a log-normal martingale in P_{j+1} by (1.15). In particular we have

$$
\begin{aligned}
L_j(T_j) &= L_j(t)\exp\left(-\frac{1}{2}\int_t^{T_j}|\gamma_j|^2(s)ds + \int_t^{T_j}\gamma_j(s)^\top dW^{(j+1)}(s)\right) \\
&\stackrel{dist.}{=} L_j(t)\exp\left(-\frac{1}{2}\int_t^{T_j}|\gamma_j|^2(s)ds + \sigma_j^B\sqrt{T_j-t}\,\zeta\right),
\end{aligned}
$$

for a real valued standard normally distributed ζ. Now the statement follows from (1.31) by elementary calculus. $\square$

Formula (1.31), essentially due to Black (1976), is usually called the Black (76) formula.

1.3.2   Swaptions in a Swap Market Model

A European (payer) swaption over period a $[T_p, T_q]$ gives the right to enter at T_p into an interest rate swap with strike rate κ, as introduced in Section 1.2.2. Clearly, the swaption contract is equivalent to a (degree 1 homogeneous) cash-flow at T_p which is equal to $B_p(T_p)\left(Swap_{p,q;\kappa}(T_p)\right)^+$. Hence the swaption value at time $t \le T_p$ is given by

$$
\begin{aligned}
Swpn_{p,q;\kappa}(t) &= B_{p,q}(t)E_{p,q}^{\mathcal{J}_t}\frac{(Swap_{p,q;\kappa}(T_p))^+}{B_{p,q}(T_p)} \\
&= B_{p,q}(t)E_{p,q}^{\mathcal{J}_t}\left(S_{p,q}(T_p)-\kappa\right)^+;
\end{aligned}
\tag{1.32}
$$

see (1.23) and (1.25).

In a swap market model the swap rate volatility is deterministic. As a consequence, in a SMM the swap rate $S_{p,q}$ is a log-normal martingale and, similar to Proposition 1.4, swaptions can be priced analytically by Black-Scholes type formulas. Indeed, if $t \to \sigma_{p,q}$ is deterministic, we may write by (1.28) for $t \le T_p$,

$$
S_{p,q}(T_p) = S_{p,q}(t)\exp\left(-\frac{1}{2}\int_t^{T_p}|\sigma_{p,q}|^2(s)ds + \int_t^{T_p}\sigma_{p,q}(s)^\top dW^{p,q}(s)\right)
$$

and obtain from (1.32) the so called Black swaption formula,

$$
Swpn_{p,q;\kappa}(t) = B_{p,q}(t)\left(S_{p,q}(t)\mathcal{N}(d_+) - \kappa\mathcal{N}(d_-)\right),\ \text{with}
\tag{1.33}
$$

$$
d_\pm := \frac{\ln\left(S_{p,q}(t)/K\right) \pm (\sigma_{p,q}^B)^2(T_p-t)/2}{\sigma_{p,q}^B\sqrt{T_p-t}},\quad (\sigma_{p,q}^B)^2 = \frac{1}{T_p-t}\int_t^{T_p}|\sigma_{p,q}|^2(s)ds.
$$

1.3.3   Approximating Swaptions in a Libor Market Model

The dynamics of swap rates in a Libor market model is rather complicated due to the stochastic factors behind the γ_l in (1.29). Therefore, closed form pricing of swaptions in a Libor market model is in general not possible. But, it is nonetheless possible to give surprisingly accurate swaption approximation formulas in an LMM.

Using (1.29) and a little algebra we may express for fixed p and q the squared magnitude of the swap volatility by

$$|\sigma_{p,q}|^2 = \frac{1}{S_{p,q}^2} \sum_{l=p}^{q-1} \sum_{l'=p}^{q-1} \gamma_l^\top \gamma_{l'} L_l L_{l'} v_l^{p,q} v_{l'}^{p,q}, \qquad \text{with} \qquad (1.34)$$

$$v_l^{p,q} = \frac{\delta_l}{1+\delta_l L_l} \left(\sum_{j=j_l}^{k-1} w_j^{p,q} \frac{B_p - B_q}{B_{p,q}} + \frac{B_q}{B_{p,q}} \right), \qquad j_l = \min\{j : p_{j+1} \geq l+1\}. \tag{1.35}$$

Let us write the possibly non-standard swap rate $S_{p,q}(t)$, $t \leq T_p$, as a sum of forward Libors,

$$S_{p,q}(t) = \frac{B_p(t) - B_q(t)}{B_{p,q}(t)} = \sum_{j=p}^{q-1} \frac{B_j(t) - B_{j+1}(t)}{B_{p,q}(t)} = \sum_{j=p}^{q-1} \widehat{v}_j^{p,q}(t) L_j(t), \tag{1.36}$$

where $B_{p,q}$ is the annuity numeraire in (1.24) or (1.26) and $\widehat{v}_j^{p,q} = \delta_j B_{j+1}/B_{p,q}$. In the case of a standard swap we have $\sum_{j=p}^{q-1} \widehat{v}_j^{p,q} = 1$, so then the coefficients $v_j^{p,q}$ can be seen as a weights. We now derive a relation between $\widehat{v}$ in (1.36) and v in (1.34). Consider the system of (virtual) swap rates $S_{l,q}^0$, for $l = p, ..., q-1$, defined by

$$S_{l,q}^0 = \frac{B_l - B_q}{B_{l,q}}, \qquad \text{with} \qquad (1.37)$$

$$B_{l,q} := \sum_{j=j_l}^{k-1} (T_{p_{j+1}} - T_{p_j}) B_{p_{j+1}} = B_{p,q} \sum_{j=j_l}^{k-1} w_j^{p,q}.$$

Note that $S_{l,q}^0$ can be seen as the swap rate over period $[T_l, T_q]$ with settlement dates $T_{j_l}, \ldots, T_{p_k} = T_q$. By (1.35) and (1.37) it easily follows that

$$v_l^{p,q} - \widehat{v}_l^{p,q} = \frac{\delta_l}{1+\delta_l L_l} \left(\sum_{j=j_l}^{k-1} w_j^{p,q} \frac{B_p - B_q}{B_{p,q}} + \frac{B_q}{B_{p,q}} \right) - \frac{\delta_l B_{l+1}}{B_{p,q}}$$

$$= \frac{\delta_l B_{l,q}}{B_{p,q}} \frac{S_{p,q}^0 - S_{l,q}^0}{1+\delta_l L_l} =: \widehat{y}_l^{p,q}, \qquad p \leq l < q. \tag{1.38}$$

The $\widehat{y}$ terms in (1.38) have magnitudes comparable with differences of swap rates; hence they are usually rather small. Moreover we have $\widehat{y}_l^{p,q} = 0$ when

$S_{l,q}^0 = S_{p,q}$ for $p < l < q$. For example, this is the case for standard swaptions when the yield curve is flat (see also Jäckel & Rebonato 2003).

Let us integrate (1.34) to obtain

$$\frac{1}{T_p - t} \int_t^{T_p} |\sigma_{p,q}|^2(s)ds = \sum_{l,l'=p}^{q-1} \int_t^{T_p} \frac{v_l^{p,q}(s)v_{l'}^{p,q}(s)L_l(s)L_{l'}(s)}{S_{p,q}^2(s)} \gamma_l(s)^\top \gamma_{l'}(s)ds,$$

(1.39)

where $t \leq T_p$ denotes the present calendar date. We now note that the (stochastic) fractions in the r.h.s. integrands of (1.39), which by (1.36), (1.38) add up to approximately one and thus may be regarded as weights, tend to vary relatively slow in practice and therefore may be approximated by their values at t. Under this additional assumption instantaneous swap volatilities may be considered as deterministic (though model inconsistent). As a result the quantities in the l.h.s. of (1.39) may be seen as squares of implied Black volatilities of approximative swaption prices in the Libor model and we thus obtain for a swaption with unit nominal and strike κ the following approximation formula.

General swaption approximation formula

$$Swpn_{p,q;\kappa}(t) \approx B_{p,q}(t)\left(S_{p,q}(t)\mathcal{N}(d_+) - \kappa\mathcal{N}(d_-)\right),$$

$$d_\pm = \frac{\ln\left(S_{p,q}(t)/\kappa\right) \pm (\sigma_{p,q}^B)^2(T_p - t)/2}{\sigma_{p,q}^B\sqrt{T_p - t}}, \quad \text{where}$$

$$(\sigma_{p,q}^B)^2 = \sum_{l,l'=p}^{q-1} \frac{v_l^{p,q}(t)v_{l'}^{p,q}(t)L_l(t)L_{l'}(t)}{S_{p,q}^2(t)} \int_t^{T_p} \gamma_l(s)^\top \gamma_{l'}(s)ds,$$

$$\text{with} \quad v_l^{p,q} = \frac{\delta_l B_{l+1}}{B_{p,q}} + \widehat{y}_l^{p,q}.$$

(1.40)

The $\widehat{y}$-term can be computed by (1.38). By the formula (1.40) we can evaluate quasi-analytically approximative model swaption prices which should be computed otherwise by tedious Monte Carlo simulation. This is very important for calibration procedures where the market model is calibrated by fitting the volatilities (1.40) directly to At-The-Money swaption volatilities quoted in the market.[1]

For our applications we highlight the following versions of the general swaption approximation formula (1.40).

- **Simple swaption approximation**

$$\widehat{y}_l^{p,q} \equiv 0$$

(1.41)

---

[1] For ATM instruments calibrating to volatilities or to prices makes in practice hardly any difference.

in (1.40) has the following important features.

- ○ *For a flat yield curve and equally settled Libors and swaps this approximation coincides with the general approximation (1.40).*

- ○ *In the simple approximation the Black volatilities in (1.40) are the same for all swap settlement conventions, since in the expression for the Black volatility in (1.40) the annuity numeraire $B_{p,q}$ cancels out if $\widehat{y} = 0$. Note, however, swap prices and the quality of approximation are generally not the same for all swap conventions.*

- **Refined swaption approximation in different markets**
  Let m be an integer, p, q such that $q - p \mod m = 0$, and set $p_j := p + mj$. It then follows that $j_l = [(l-p)/m]$, with $[x] := \max\{k \in \mathbb{Z} : k \leq x\}$ denoting the Entier function, and obtain for the correction term (1.38) by a little algebra,

$$\widehat{y}_l^{p,q} = \frac{\delta_l}{B_{p,q}(1 + \delta_l L_l)} \left(B_p - B_l - S_{p,q} \sum_{j=0}^{[\frac{l-p}{m}]-1} B_{p+m(j+1)} \sum_{r=p}^{p+m-1} \delta_{mj+r} \right).$$

(1.42)

Thus, (1.42) yields refined swaption approximation formulas for application in different markets:

- ○ $m = 1$ for the standard (equally settled) swaption market,

- ○ $m = 2$ for Euro swaptions in a semi-annual Libor model,

- ○ $m = 4$ for US, UK, and Japanese swaptions in a quarterly Libor model.

The fact that for standard swaptions the simple swaption approximation (1.40), (1.41) and the refined approximation (1.40), (1.42) coincide when the initial yield curve is flat was pointed out and explained previously by Jäckel & Rebonato (2003). For a typical flat yield curve and LMM volatility structure the relative errors between approximated swaption prices due to (1.40), (1.41) and Monte Carlo simulated prices turned out to be about 0.3%. For a non-flat yield curve and refined formula (1.40), (1.42) with $m = 1$, the relative errors turned out to be 0.3% as well relative in this paper.

The conclusions of Jäckel & Rebonato (2003) were confirmed by our simulation tests. But, we then considered annually settled swaptions in a semi-annual Libor model, for instance, for modelling an EurIbor environment. By comparing (accurate enough) Monte Carlo prices with approximations due to (1.40), (1.41) for a flat yield curve, our experiments showed that the accuracy of the simple approximation (1.40), (1.41) was significantly less, namely, about 1% relative for a flat yield curve and about 2% relative for some typical non-flat curve. Of course this is due to the fact that the $\widehat{y}$ term in (1.40) not vanishes for non-standard swaptions in general. However, by applying the refined formula (1.40), (1.42) with $m = 2$ we obtained relative errors of about 0.3% again. We therefore draw the following conclusion.

For differently settled caps and swap(tion)s, in particular in the Euro, US, UK, and Japanese market, we recommend to use the refined swaption formula (1.40), (1.42) for a suitable m in any case, whether the initial yield curve is flat or not.

1.3.4 Smile/Skew Extensions of the Libor Market Model

We have seen that a caplet can be priced by Black's formula (1.31) in a LMM. Conversely, given a caplet price, strike, and maturity, one can ask for the so called *implied caplet volatility* such that (1.31) holds. In the markets, caps, caplets, and swaptions are usually quoted by (implied) volatilities rather than prices. Obviously, the implied Black volatility of a caplet price due to a Libor market model is flat with respect to caplet strike and maturity. This is actually not consistent with reality, where usually caplet volatilities as function of the strike behave like a skew, or an asymmetric smile.

An obvious way to allow for smile effects in the Libor model is to consider in (1.22) instead of a deterministic volatility structure a more general volatility process of the form, $t \to \gamma_i(t, L(t))$, where for $1 \le i < n$, $\gamma_i(\cdot, \cdot)$ is a certain deterministic $\mathbb{R}^m$-valued vector function which satisfies suitable conditions. If we take $\gamma_i(t, L) = \gamma_i^{(0)}(t) L_i^{\alpha - 1}$, where $\alpha > 0$ and $t \to \gamma_i^{(0)}(t) \in \mathbb{R}^m$ is deterministic, we have the so called CEV model which allows for closed form caplet prices. For details see Schroder (1989) and Andersen & Andreasen (2000), where a finite difference procedure for a volatility structure of the form $\gamma_i(t, L) = \gamma^{(0)}(t)\phi(L_i)$, $1 \le i < n$, is proposed. As a more general approach to cover smile effects in model prices of caps and swaptions, one considers in the literature Libor models driven by Lévy processes rather than Brownian motions; see Eberlein & Özkan (2002), Glasserman & Kou (2003), Glasserman & Merener (2002), and Jamshidian (1999).

While extensions of the Libor market model give more flexibility and thus may reflect reality more properly, the pricing of caps and swaptions is generally more complicated and time consuming in such models. This may be a problem when these pricing procedures are used in a procedure of calibrating to caps and swaptions. Further, in particular for jump-diffusion Libor models, the larger model flexibility may lead to serious instability of calibration procedures. For a treatment of stability problems connected with the identification of Lévy based models, such as jump-diffusions, we refer to Cont & Tankov (2003). In this book we restrict ourselves to the standard Libor market model and its calibration issue, although we do not want to say that more general Libor models are superfluous luxury. For instance, it will be certainly possible to extend the calibration procedures for a parametric Libor model proposed in Chapter 3 to certain generalized Libor models, for example the CEV model.

Chapter 2

Parametrisation of the Libor Market Model

In this chapter we study different parametrisations of the Libor market model. After some general considerations we discuss the issue of (quasi) time-shift homogeneity and introduce correlation functions and correlation structures in this context. In particular, we present a systematic approach for deriving low parametric correlation structures with nice properties, which allow for efficient and stable calibration procedures developed in Chapter 3.

2.1 General Volatility Structures

From now on we consider the Libor Market Model with respect to a tenor structure $0 < T_1 < T_2 < \ldots < T_n$ in the *spot Libor measure* P^*,

$$dL_i = \sum_{j=m(t)}^{i} \frac{\delta_j L_i L_j}{1 + \delta_j L_j} \gamma_i^\top \gamma_j \, dt + L_i \, \gamma_i^\top dW^*, \qquad (2.1)$$

where $(W^*(t) \mid 0 \leq t < T_{n-1})$ is a standard d-dimensional Wiener process under the measure P^* with d, $1 \leq d < n$, being the number of driving factors and $m(t) := \min\{m : T_m \geq t\}$ denoting the next reset date at time t. See for details Section 1.2.1

For convenience we define the *absolute or scalar volatility process* σ of the forward Libor rate L_i by

$$t \to \sigma_i(t) := |\gamma_i(t)| = \sqrt{\sum_{k=1}^{d} \gamma_{i,k}^2(t)}, \qquad 0 \leq t \leq T_i, \ 1 \leq i < n, \qquad (2.2)$$

and *correlation structure* ρ by

$$t \to \rho_{ij}(t) := \frac{\gamma_i(t)^\top \gamma_j(t)}{|\gamma_i(t)||\gamma_j(t)|}, \qquad 0 \leq t \leq \min(T_i, T_j), \ 1 \leq i,j < n. \qquad (2.3)$$

It should be noted that the terms volatility and correlation here should be interpreted as local or instantaneous volatility and correlation. It is clear that for any deterministic possibly time dependent d-dimensional orthogonal matrix function Q, i.e., $Q^\top Q = QQ^\top = I_d$ with I_d being the d-dimensional identity matrix, the volatility structure $\widetilde{\gamma}$ defined by

$$\widetilde{\gamma}_{i,k} := \sum_{l=1}^{d} \gamma_{i,l} Q_{lk} \tag{2.4}$$

yields the same scalar volatility and correlation structure via (2.2), (2.3). So the volatility structure γ is not completely determined by specification of the structures σ and ρ alone. However, we have the following proposition.

PROPOSITION 2.1
Consider a deterministic matrix valued map

$$Q : [0, T_{n-1}] \ni t \to Q(t) \in \mathbb{R}^{d \times d},$$

such that for each t the matrix $Q(t)$ is orthogonal, i.e., $Q^\top Q = QQ^\top = I_d$. Let us assume further the weak restriction that Q is Lebesgue measurable and bounded. Then, if the process W is a d-dimensional standard Brownian motion, $\widetilde{W}$ defined by

$$\widetilde{W}_t = \int_0^t Q(u) dW_u$$

is again a d-dimensional standard Brownian motion.

PROOF Since a zero-mean Gaussian process is uniquely determined by its covariance function, this follows from the fact that $\widetilde{W}$ is a Gaussian process with $E\widetilde{W}_t \equiv 0$, and

$$E\widetilde{W}_t \widetilde{W}_s^\top = E \int_0^t Q(u) dW_u \left(\int_0^s Q(u) dW_u \right)^\top$$
$$= \int_0^{\min(t,s)} I_d du = \min(t, s) I_d,$$

which is the covariance function of d-dimensional standard Brownian motion.
▯

COROLLARY 2.1
*If two volatility structures γ and $\widetilde{\gamma}$ are related by (2.4), their corresponding scalar volatility and correlation structures coincide and their corresponding Libor processes are identical in distribution. In this sense, γ and $\widetilde{\gamma}$ may be regarded as **equivalent** volatility structures.*

Economically, only the distribution of the Libor process is relevant. Therefore, by Proposition 2.1 and Corollary 2.1, the scalar volatility structure σ

and correlation structure ρ are the basic economical objects which determine the Libor market model and, loosely speaking, given these objects one may choose (almost) any convenient deterministic volatility structure γ such that (2.2) and (2.3) hold.

2.2   (Quasi) Time-Shift Homogeneous Models

2.2.1   Correlation Structures from Correlation Functions

From an economical point of view it makes sense to have some sort of time shift homogeneity in the Libor market model. Therefore it is natural to consider scalar volatility and correlation structures of the following special type,

$$\sigma_i(t) := c_i g(T_i - t), \qquad 0 \le t \le T_i, \quad c_i > 0,$$
$$\rho_{ij}(t) := \varrho(T_i - t, T_j - t), \quad 0 \le t \le \min(T_i, T_j), \quad 1 \le i, j < n, \quad (2.5)$$

for a non-negative function $g : [0, \infty) \ni u \to \mathbb{R}_+$ and a function $\varrho : [0, \infty) \times [0, \infty) \ni (u, v) \to [-1, 1]$, which satisfies the following conditions:

(i) $\varrho(u, v) = \varrho(v, u)$, for $u, v \ge 0$ (symmetry),

(ii) $\varrho(u, u) = 1$, $|\varrho(u, v)| \le 1$, for $u, v \ge 0$,

(iii) For any number $K \in \mathbb{N}$, any set $\{u_1, \ldots, u_K\} \subset \mathbb{R}_+^K$ and $\{v_1, \ldots, v_K\} \subset \mathbb{R}^K$, it holds (non-negativity)

$$\sum_{k,l=1}^{M} \varrho(u_k, u_l) v_k v_l \ge 0.$$

A function ϱ which satisfies (i), (ii), and (iii), is called a *correlation function*.

If in (2.5) the volatility coefficients c_i are taken to be identical, i.e., $c_i = c_1$ for all i, we say that volatility structure as well as the corresponding Libor market model is *time shift homogeneous*. If $c_i \not\equiv c_1$ we speak of a quasi time-shift homogeneous structure and model. The reason for including the coefficients c_i is that in certain applications pure time shift homogeneity of the market model may be too restrictive for calibration purposes, as we will discuss in Chapter 3.

2.2.2   Finitely Decomposable Correlation Functions

Let us consider for an integer d a unit-vector function $\eta : [0, \infty) \ni u \to \mathbb{R}^d$, hence $|\eta(\cdot)| \equiv 1$. Such a function is easily obtained by normalizing some arbitrary $\mathbb{R}^d$-vector function. Then, η defines a correlation function via

$$\varrho(u, v) = \eta^{\top}(u)\,\eta(v). \tag{2.6}$$

It is clear that for any sequence $0 < u_1 < \cdots < u_q$, $[\rho(u_i, u_j)]_{1 \le i, j \le q}$ is a $q \times q$ correlation matrix with rank not larger than $\min(q, d)$. We will say that a correlation function which satisfies (2.6) for some d and η is *finitely decomposable*.

If for a finitely decomposable correlation function d and η are known, we can write down immediately a d-factor volatility structure consistent with (2.5):

$$\gamma_{i,k}(t) = c_i g(T_i - t)\eta(T_i - t), \qquad 1 \le i < n,\ 1 \le k \le d. \tag{2.7}$$

REMARK 2.1   The class of finitely decomposable correlation functions is rather special as, for instance, it is not difficult to see that the correlation function $\varrho(u, v) := \exp(-|u - v|)$ does not belong to this class. ⬚

2.2.3   Ratio Correlation Structures and Functions

The following proposition, due to Curnow and Dunnet (1962) provides a basis for systematic generation of economically sensible correlation functions and structures for the Libor market model.

PROPOSITION 2.2
Let (b_k), $k = 1, \ldots, K$, be an arbitrary non-decreasing sequence. Then the matrix ρ defined by

$$\rho_{kl} := \frac{b_k}{b_l}, \qquad \rho_{lk} := \rho_{kl}, \qquad 1 \le k \le l \le K, \tag{2.8}$$

is a correlation matrix. Moreover, we have a Cholesky decomposition $\rho = \eta\eta^{\top}$ with

$$\eta_{kl} := \frac{a_l}{b_k} 1_{l \le k}, \qquad 1 \le k, l \le K$$

where the sequence (a_l) is defined by $a_l := \sqrt{b_l^2 - b_{l-1}^2}$, $1 \le l \le K$ and $b_0 := 0$. If, in addition, the sequence (b_k) is strictly increasing, the correlation structure (2.8) has full rank.

PROOF   Let Z_p, $p = 1, \ldots, K$, be standard normally distributed independent real random variables and consider the random variables

$$Y_k := \sum_{p=1}^{k} a_p Z_p. \tag{2.9}$$

Then, for $k \leq l$ the covariance between Y_k and Y_l is given by

$$Cov(Y_k, Y_l) = \sum_{p=1}^{k} a_p^2 = b_k^2$$

and so the correlation is given by $Cor(Y_k, Y_l) = b_k/b_l$. Hence, (2.8) defines a correlation matrix indeed. The Cholesky decomposition can be verified straightforwardly. If (b_k) is strictly increasing we have $\eta_{kk} > 0$ for $1 \leq k \leq K$. So then the lower triangular matrix η has full rank and thus ρ also. $\qquad \square$

If we have a non-monotonic positive sequence b, we can first rearrange b to a non-decreasing sequence $\tilde{b}$ by a permutation κ, $\tilde{b}_j := b_{\kappa(j)}$, and then consider the correlation matrix $\tilde{\rho}$ obtained by $\tilde{b}$ via (2.8). Hence, for $1 \leq k, l \leq K$ we have the correlation matrix $\tilde{\rho}_{kl} = \min(\tilde{b}_k, \tilde{b}_l)/\max(\tilde{b}_k, \tilde{b}_l)$, and so $\rho_{kl} := \tilde{\rho}_{\kappa^{-1}(k), \kappa^{-1}(l)} = \min(b_k, b_l)/\max(b_k, b_l)$ is a correlation matrix as well.

COROLLARY 2.2

For any possibly non-monotonic positive sequence (b_k) we have that ρ defined by

$$\rho := \frac{\min(b_k, b_l)}{\max(b_k, b_l)}, \qquad 1 \leq k, l \leq K,$$

is a correlation matrix.

COROLLARY 2.3

Consider a fixed but arbitrary positive function $\beta : [0, \infty) \to \mathbb{R}_{++}$. Then, the function

$$\varrho(u, v) = \frac{\min(\beta(u), \beta(v))}{\max(\beta(u), \beta(v))}, \qquad u, v \geq 0, \qquad (2.10)$$

is a correlation function.

PROOF   The conditions *(i)* and *(ii)* are trivially satisfied. To prove the non-negativity condition *(iii)*, take an arbitrary $K \in \mathbb{N}$ and a set of non-negative numbers $\{u_1, \ldots, u_K\}$. Then apply Corollary 2.2 to the sequence (b_k) with $b_k := \beta(u_k)$ for $1 \leq k \leq K$. $\qquad \square$

Let us now take an *increasing* function $\beta : [0, \infty) \to \mathbb{R}_{++}$ and consider the correlation structure (2.5) for the correlation function (2.10). Hence,

$$\varrho_{ij}(t) := \frac{\min[\beta(T_i - t), \beta(T_j - t)]}{\max[\beta(T_i - t), \beta(T_j - t)]}, \qquad t \leq \min(T_i, T_j), \quad 1 \leq i, j < n. \quad (2.11)$$

Since β is increasing, the following decomposition is clear from Proposition 2.2. For fixed t, $0 \le t \le T_{n-1}$, we have

$$\varrho_{ij}(t) = \sum_{k=m(t)}^{n-1} \eta_{ik}(t)\eta_{jk}(t), \quad m(t) \le i,j < n,$$

with

$$\eta_{ik}(t) := \frac{\sqrt{\beta^2(T_k - t) - 1_{m(t)<k}\beta^2(T_{k-1} - t)}}{\beta(T_i - t)} 1_{m(t) \le k \le i}.$$

We thus obtain a Libor market model (2.1) consistent with (2.5) for correlation structure (2.11) by taking $d = n - 1$ and volatility structure

$$\gamma_{ik}(t) = c_i g(T_i - t)\eta_{ik}(t), \qquad m(t) \le i < n,\ 1 \le k < n. \tag{2.12}$$

REMARK 2.2
(i) Note that in general the correlation function (2.10) is not finitely decomposable in the sense of (2.6). Therefore, the decomposing vector function $\eta_i := (\eta_{ik})_{1 \le k < n}^\top$ in (2.12) will in general not be time shift homogeneous, i.e., of the form $\eta(T_i - t)$ as in (2.7).
(ii) Since the function β in (2.11) is assumed to be strictly increasing, the correlation matrix $\varrho(t)$ has rank $n - 1$ for $t < T_1$. Therefore we need in (2.1) an $(n - 1)$-dimensional Brownian motion, i.e., we need to take $d = n - 1$.
(iii) The increasingness of β has a sound economical interpretation. Namely, that the (instantaneous) correlation between forward Libors L_i and L_{i+k} *decreases* with k.   ☐

2.2.4   LMM with Piece-wise Constant Volatility Structure

The Hull-White Parametrisation

In Hull and White (2000) a parametrisation of the Libor market model is proposed where basically the scalar volatility and correlation functions only depend on the number of reset dates between the actual time and maturities of forward Libors.

Let $\rho^{(0)} := [\rho_{ij}^{(0)}]_{0 \le i,j < n-1}$ be a fixed correlation matrix of rank d and $\Lambda := [\Lambda_i]_{0 \le i < n-1}$ be a fixed vector with non-negative entries. Then the Hull-White parametrisation comes down to the following choice of scalar volatility and correlation functions,

$$\sigma_i(t) := \Lambda_{i-m(t)}, \qquad 0 \le t \le T_i,$$

$$\rho_{ij}(t) := \rho_{i-m(t),j-m(t)}^{(0)}, \quad 0 \le t \le \min(T_i, T_j), \quad 1 \le i,j < n. \tag{2.13}$$

Let us take any set of decomposing unit vectors $\{e_i^{(0)} \in \mathbb{R}^d : 0 \le i < n - 1\}$ with $\rho_{ij}^{(0)} = \left(e_i^{(0)}\right)^\top e_j^{(0)}$, $0 \le i,j < n - 1$, and set $e_i(t) := e_{i-m(t)}^{(0)}$, $1 \le i < n$.

Due to Proposition 2.1 and Corollary 2.1 the dynamics of the Hull-White model may be represented by SDE (2.1), with volatility structure γ given by

$$\gamma_i(t) := \sigma_i(t)e_i(t) = \Lambda_{i-m(t)}e_{i-m(t)}^{(0)}. \tag{2.14}$$

REMARK 2.3 For an *equidistant* tenor structure, i.e., $\delta_i = \delta_1 =: \delta$ for $1 \leq i < n$, the Hull-White volatility structure is via (2.2) and (2.3) consistent with a special version of the general (quasi) time-shift homogeneous volatility and correlation structure (2.5). Indeed, take the piecewise constant function g with $g(u) := \Lambda_{k-1}$ for k, $1 \leq k < n$, such that $(k-1)\delta \leq u < k\delta$, together with a correlation function ϱ, defined by $\varrho(u,v) := \rho_{k-1,l-1}^{(0)}$ for k,l, $1 \leq k,l < n$ such that $(k-1)\delta \leq u < k\delta$, $(l-1)\delta \leq v < l\delta$, and then take $c_i := 1$, $1 \leq i < n$. In particular, the Hull-White correlation function ϱ is decomposable since we have

$$\varrho(u,v) = \rho_{k-1,l-1}^{(0)} = \left(e_{k-1}^{(0)}\right)^{\top} e_{l-1}^{(0)} =: e(u)^{\top}e(v).$$

▯

Caplet pricing

From the structure of the Hull-White model it follows that the system of Black volatilities $\{\sigma_i^B : 1 \leq i < n\}$ consistent with model prices of caplets satisfies

$$(\sigma_i^B)^2 T_i = \int_0^{T_i} \sigma_i^2(t)dt$$

$$= \sum_{k=1}^{i} \Lambda_{i-k}^2 \delta_{k-1} = \sum_{k=0}^{i-1} \Lambda_k^2 \delta_{i-k-1}, \quad 1 \leq i < n, \tag{2.15}$$

with $\delta_0 := T_1$. In fact, the relationship (2.15) opens up the possibility of perfectly calibrating the scalar vector Λ to a system of market (for instance ATM) caplet volatilities, say $\{\hat{\sigma}_i^B : 1 \leq i < n\}$. Indeed, we can construct the vector Λ by induction: For $i = 1$ in (2.15) we get $\Lambda_0 = \hat{\sigma}_1^B$, and, if Λ_i, $0 \leq i < m$ are known, we obtain Λ_m by taking $i = m + 1$ in (2.15),

$$(\hat{\sigma}_{m+1}^B)^2 T_{m+1} - \sum_{k=0}^{m-1} \Lambda_k^2 \delta_{m-k} = \Lambda_m^2 \delta_0. \tag{2.16}$$

REMARK 2.4 It is important to note that due to (2.16) exact caplet fitting in the Hull-White model is only possible when the market caplet volatilities are such that

$$(\hat{\sigma}_{m+1}^B)^2 T_{m+1} - \sum_{k=0}^{m-1} \Lambda_k^2 \delta_{m-k} \geq 0, \tag{2.17}$$

for $1 \leq m < n - 1$. For an equidistant tenor structure this comes down to

$$(\hat{\sigma}_{m+1}^B)^2 T_{m+1} - (\hat{\sigma}_m^B)^2 T_m \geq 0, \qquad 1 \leq m < n - 1. \tag{2.18}$$

So for an exact fit the caplet volatilities may not decrease too fast with maturity in fact. Unfortunately, usually (2.17) and respectively (2.18) are violated
in practice; see Section 3.5. Thus, in practice perfect matching of caplets with
the Hull-White model is usually not possible. $\quad\quad\quad\quad\quad\quad\quad\quad\quad\quad\quad$ □

2.2.5   Modified Hull-White Volatility Structure

The Hull-White volatility structure (2.14) has the following important features: *(i)* The structure is pure time shift homogeneous when the tenor structure is equidistant (see Remark 2.3). *(ii)* For a non-equidistant but not too
irregular tenor structure the volatilities can still be regarded as "nearly" time
shift homogeneous. *(iii)* The pure or "nearly" time shift homogeneous correlation structure can be decomposed canonically into volatility factors which
are also pure or "nearly" time shift homogeneous and vice versa. However, as
noted in Remark 2.4 and illustrated in Section 3.5, a drawback of this model
is the breakdown of recursive calibration of Λ's to caplets. On the other hand,
while exact caplet fitting of the Λ's is usually impossible, a least squares calibration of Λ's is likely to be unstable due to the large number of fitting
parameters, e.g., $n \approx 40$. Therefore, instead of Λ's one could take a scalar
volatility structure as in (2.5), where the function g is (parsimoniously) parametrized in some suitable way. An important benefit of such scalar volatility
structure is the possibility of using a smooth function g, hence smooth scalar
volatility $|\gamma_i(t)|$, for instance, by choosing a nice smooth functional form of g,
depending on some set of parameters. Of course, while smoothness of volatility norms can thus be retained, there is some price we have to pay. Namely,
we need to include the coefficients c_i which are necessary to match the caplet
vols $\hat{\sigma}_i^B$ via

$$(\hat{\sigma}_i^B)^2 T_i = \int_0^{T_i} |\gamma_i(t)|^2 dt = c_i^2 \int_0^{T_i} g^2(s) ds. \tag{2.19}$$

Note that for any particular choice of g the caplets may be matched perfectly
by choosing c_i according to (2.19). However, some arbitrary choice of g would
take the volatility structure far from time shift homogeneity, which might be
economically less desirable. Therefore, if one would like to keep the model as
close to time shift homogeneity as possible, a way to go would be, for instance,
determination of a function g in some nice parametric class $\{g_\alpha : \alpha \in A\}$, such
that

$$\sum_{i=1}^{n-1} \left((\hat{\sigma}_i^B)^2 T_i - c^2 \int_0^{T_i} g_\alpha^2(s) ds \right)^2 \to \min_{\alpha \in A, \, c > 0}, \tag{2.20}$$

and then determine c_i by (2.19).

The considerations above suggest to combine the Hull-White structure (2.14) with the structure (2.5) into the alternative volatility structure,

$$\gamma_i(t) = c_i g(T_i - t) e_{i-m(t)}^{(0)}, \qquad 0 \leq t \leq \min(T_i, T_j), \quad 1 \leq i, j < n,$$

$$\rho_{kl}^{(0)} = \left(e_k^{(0)} \right)^\top e_l^{(0)}, \qquad\qquad 0 \leq k, l < n - 1. \tag{2.21}$$

Thus, the structure (2.21), which can be seen as a Modified Hull-White volatility structure, contains the scalar volatility structure from (2.5) and the correlation structure from (2.14). In fact, the volatility structure (2.21) equipped with a suitably parametrised volatility function g (see (2.25) in Section 2.2.6) and a correlation matrix $\rho^{(0)}$ as developed in Section 2.3.4 will be our favorite structure for applications.

REMARK 2.5 *Non-standard forward Libors*

In practice it may happen that one needs to price a product for which the underlying forward rates do not fit with the presently standard tenor structure. For such a situation we here give a simple pragmatic way for modelling such non-standard forward Libors similar as in Brigo & Mercurio (2001a).

Note that from (1.18) we have

$$dW^{(i+1)} = dW^{(i)} + \frac{\delta_i L_i}{1 + \delta_i L_i} \gamma_i dt, \tag{2.22}$$

where $W^{(i+1)}$ and $W^{(i)}$ are standard Brownian motions under P_{i+1} and P_i, respectively. Consider the non-standard forward Libor $L_{i-1}^{\alpha,\beta}$ with respect to forward period $[T_{i-1}^\alpha, T_i^\beta]$, where $T_{i-1}^\alpha := (1 - \alpha)T_{i-1} + \alpha T_i$, $T_i^\beta := (1 - \beta)T_i + \beta T_{i+1}$, and $0 \leq \alpha, \beta \leq 1$. We next introduce an intermediate standard Brownian motion $W_\beta^{(i)}$ under an intermediate measure P_i^β, connected with expiry T_i^β, by a drift interpolation,

$$dW_\beta^{(i)} := dW^{(i)} + \frac{(T_i^\beta - T_i)L_i}{1 + (T_i^\beta - T_i)L_i} \gamma_i dt. \tag{2.23}$$

Note that $\beta = 1$ yields (2.22) and $\beta = 0$ an identity. Then, for the dynamics of $L_{i-1}^{\alpha,\beta}$, where $\alpha \neq 0 \vee \beta \neq 1$, we postulate

$$dL_{i-1}^{\alpha,\beta} = L_{i-1}^{\alpha,\beta} \left(\gamma_{i-1}^\alpha \right)^\top dW_\beta^{(i)} \tag{2.24}$$

$$= \frac{(T_i^\beta - T_i)L_i L_{i-1}^{\alpha,\beta}}{1 + (T_i^\beta - T_i)L_i} \left(\gamma_{i-1}^\alpha \right)^\top \gamma_i dt + L_{i-1}^{\alpha,\beta} \left(\gamma_{i-1}^\alpha \right)^\top dW^{(i)},$$

where the intermediate deterministic volatility γ_{i-1}^α is given by

$$\gamma_{i-1}^\alpha(t) := ((1 - \alpha)c_{i-1} + \alpha c_i)g((1 - \alpha)T_{i-1} + \alpha T_i - t)e_{i-1-m(t)}^{(0)},$$
$$0 \leq \alpha < 1, \qquad 2 \leq i < n,$$

and thus extends structure (2.21). In this approach, two non-standard Libors $L_{i-1}^{\alpha,\beta}$ and $L_{j-1}^{\alpha_1,\beta_1}$, with $0 \leq \alpha, \beta \leq 1$ and $0 \leq \alpha_1, \beta_1 \leq 1$, are perfectly correlated for $i = j$. Finally we note that the interpolation (2.23) differs slightly from the interpolation in Brigo & Mercurio (2001a) with respect to the denominator in the second term. In fact, by interpolation (2.23) it is guaranteed that the by (2.24) extended Libor model is automatically arbitrage-free.   ⬜

2.2.6   Parametric Scalar Volatility Function

Particularly for calibration purposes, we prefer to use in (2.21) a scalar volatility function g proposed by Rebonato (1999), which is of the following parametric form,

$$g(s) = g_{a,b,g_\infty}(s) := g_\infty + (1 - g_\infty + as)e^{-bs}, \quad a, b, g_\infty > 0. \qquad (2.25)$$

In fact, (2.25) is obtained as a constant plus a linear combination of the first two (orthogonal) Laguerre functions $e^{-s/2}$ and $(s-1)e^{-s/2}$ in $[0, \infty)$, properly scaled.

From a modelling point of view, the parametric function g_{a,b,g_∞} is able to reflect frequently observed "hump shape" behaviour of forward Libor volatilities, before reaching their maturity. See Figure 2.1 for a typical example.

Technically, (2.25) has the advantage that integrals of the form

$$\int_0^{T_k} g_{a,b,g_\infty}(T_i - s)g_{a,b,g_\infty}(T_j - s)ds, \qquad T_k \leq \min(T_i, T_j), \qquad (2.26)$$

which will play an important role in the approximation of swaptions, can be evaluated in closed form.

REMARK 2.6   The explicit expression for the integral (2.26) is horribly long and for this reason not given here. However, such an expression can be easily produced with a formula processor like Derive, Maple, or Mathematica, and then, for instance, the result can be copied into some relevant C++ implementation.   ⬜

2.3   Parametrisation of Correlation Structures

2.3.1   A Disadvantage of Low Factor Models

It is known that low factor models have intrinsic problems to match (instantaneous) correlations between forward Libors realistically, see, e.g., Rebonato

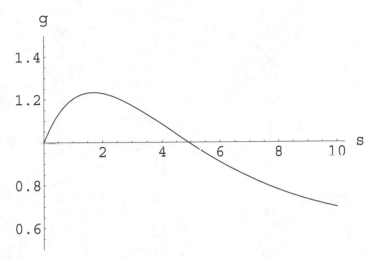

FIGURE 2.1: A typical volatility function: $s \to g_{0.5,0.4,0.6}(s)$

(1996), Schoenmakers & Coffey (1998). Let us illustrate this fact by considering a two factor Libor model with volatility structure of type (2.21), where

$$\gamma_i(t) = c_i g(T - t_i) e^{(0)}_{i-m(t)}, \quad e^{(0)}_k \subset \mathbb{R}^2, \quad 1 \le i < n, \qquad (2.27)$$

and $e^{(0)}_k$, $0 \le k < n - 1$, are unit vectors which determine the initial instantaneous correlations $\rho^{(0)}_{kl} = \left(e^{(0)}_k\right)^{\top} e^{(0)}_l$.

Let us consider the forward Libor return correlations in the first period after $t = 0$. Since the correlation matrix $\rho^{(0)}$ has rank two, we may write

$$\frac{\gamma_i(0+)^{\top}\gamma_j(0+)}{|\gamma_i(0+)||\gamma_j(0+)|} = \rho^{(0)}_{i-1,j-1} = \left(e^{(0)}_{i-1}\right)^{\top} e^{(0)}_{j-1} = \cos(\phi_{i-1} - \phi_{j-1}), \quad (2.28)$$

for some set of angles $\phi_0, \ldots, \phi_{n-2}$ with $\phi_0 := 0$.

We now consider an artificial example. Take $n = 20$ and suppose that the market tells us the correlations $\rho^{(0)}_{0,j-1}$ behave like $\rho^{(0)}_{0,j-1} = 18/(17 + j)$, thus falling down from 1 to 0.5. Then, if we calibrate this two-factor model, i.e., the ϕ_k, to these correlations it is easily seen that, as an immediate consequence, the model correlations $\rho^{(0)}_{j-1,18}$ are given by $\rho^{(0)}_{j-1,18} = \frac{9}{17+j} + \frac{\sqrt{3}}{2}\sqrt{1 - \left(\frac{17}{18} + \frac{j}{18}\right)^{-2}}$; see Figure 2.2. However, the behaviour of the correlations $\rho^{(0)}_{j-1,18}$ in figure (2.2) is clearly *not* consistent with expected real behaviour in the market which should look more or less the same as $\rho^{(0)}_{0,j-1}$, mirrored at $j = 10$. Of course the situation will be better when we increase the number of factors but a two factor example reveals this problem most clearly. In this respect we note that in Brigo and Mercurio (2001, 2001a)

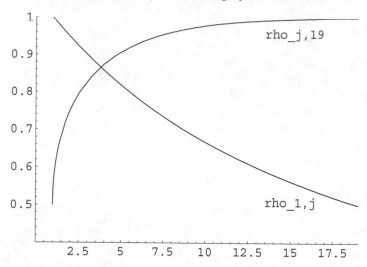

FIGURE 2.2:   Example of two factor correlations

calibration of the cosine structure (2.28) is extensively tested and their conclusions concerning the behaviour of implied correlations more or less confirm
the difficulties sketched above.

REMARK 2.7   It should be noted that the above considerations are not
affected by the particular choice of the scalar volatility function in (2.21),
since the instantaneous correlations are completely determined by the unit
vectors $e_k^{(0)}$ alone.   □

As a solution for the intrinsic low factor problem we consider in the next
section multi-factor models based on a natural framework of ratio correlation
structures (Section 2.2.3), earlier proposed in Schoenmakers & Coffey (1998)
and further developed in Schoenmakers & Coffey (2000,2003), in which correlation behaviour observed in practice is nicely reflected. Correlation structures in this framework involve basically the same number of parameters as
the general rank two correlation structure (2.28).

2.3.2   Semi-parametric Framework for Libor Correlations

We consider a multi-factor Libor market model (2.1) consistent with volatility structure (2.21) with $d = n - 1$, and where $e_i^{(0)}$ are unit vectors which
determine a particular initial correlation structure as introduced below.

A Semi-parametric Libor Correlation Structure is by definition a matrix $\rho^{(0)}$ such that

$$\rho^{(0)}_{kl} = \left(e^{(0)}_k\right)^\top e^{(0)}_l = \frac{\min(b_k, b_l)}{\max(b_k, b_l)}, \quad 0 \le k, l < n-1, \qquad (2.29)$$

for a strictly increasing sequence $b = (b_0, \dots, b_{n-2})$ which is such that in addition

$$k \longrightarrow \frac{b_k}{b_{k+1}} \text{ is strictly increasing.} \qquad (2.30)$$

Since a correlation structure (2.29) essentially involves $\mathcal{O}(n)$ parameters, whereas a general or *non-parametric* correlation matrix would require $\mathcal{O}(n^2)$ entries, we may call (2.29) a *semi-parametric* correlation structure.

The increasingness of the sequence b implies that the instantaneous model correlations $Cor(\Delta L_i, \Delta L_{i+k})$ *decrease* with k for fixed i, and the additional assumption (2.30) implies that $Cor(\Delta L_i, \Delta L_{i+k})$ *increases* with i for fixed k. Both implications reflect natural behaviour of forward rates. For instance, on the one hand, the correlation between a two and a ten year forward is expected to be lower than the correlation between a two and five year forward whereas, on the other hand, the correlation between a six and ten year forward is expected to be higher than the correlation between a two and six year forward.

Because of Corollary 2.1 we may assume without restriction that $e^{(0)}$ is lower triangular. In particular, due to the imposed correlation structure (2.29) and Proposition (2.2), we can take $e^{(0)}_{k,l} = 1_{0 \le l \le k} \frac{1}{b_k} \sqrt{b_l^2 - b_{l-1}^2}$ with $b_{-1} := 0$. Hence, with $g_i(t) := c_i g(T_i - t)$ we may write

$$\Delta L_i = \dots \Delta t + L_i g_i \sum_{l=0}^{i-m(t)} e^{(0)}_{i-m(t),l} \Delta W_l^*$$

$$= \dots \Delta t + L_i g_i \sum_{l=0}^{i-m(t)} \frac{1}{b_{i-m(t)}} \sqrt{b_l^2 - b_{l-1}^2} \Delta W_l^* \qquad (2.31)$$

$$= \dots \Delta t + L_i g_i \sqrt{1 - \frac{b_{i-m(t)-1}^2}{b_{i-m(t)}^2}} \Delta W_i^* + \frac{L_i g_i}{L_{i-1} g_{i-1}} \frac{b_{i-m(t)-1}}{b_{i-m(t)}} \Delta L_{i-1}.$$

The interpretation of (2.31) is clear: The risky part of a forward Libor increment ΔL_i at time t, $t < T_{i-1}$, is a linear combination of the forward increment ΔL_{i-1} and an independent random shock ΔW_i^*, with coefficients determined by $L_{i-1}(t)$, $L_i(t)$, $g_{i-1}(t)$, $g_i(t)$, $b_{i-m(t)-1}$, and $b_{i-m(t)}$, respectively. We emphasize that decomposition (2.31) is due to the special structure of the correlation matrix (2.29) and does not hold for a general (non-parametric) correlation structure.

2.3.3   Representation Theorems

The next theorem provides a representation for the sequence b in the semi-parametric structure (2.29), where the non-linear constraints on b are transformed into simple constraints in $\mathbb{R}_+^{n-2}$.

THEOREM 2.1

For every non-decreasing sequence $b := (b_1, \ldots, b_m)$ with $b_1 := 1$ and b such that $i \to b_i/b_{i+1}$ non-decreasing, there exists a sequence of non-negative numbers $\Delta_i, \Delta_i \geq 0$, $i = 2, \ldots, m$, such that

$$b_i = \exp\left[\sum_{l=2}^{m} \min(l-1, i-1)\Delta_l\right], \quad 1 \leq i \leq m. \qquad (2.32)$$

Conversely, for any sequence Δ_l with $\Delta_l \geq 0$, $l = 2, \ldots, m$, (2.32) is non-decreasing, b_i/b_{i+1} is non-decreasing, $b_1 = 1$. Moreover, if b is given by (2.32) then

$$\Delta_i = 2\ln b_i - \ln b_{i-1} - \ln b_{i+1}, \qquad 2 \leq i < m,$$
$$\Delta_m = \ln b_m - \ln b_{m-1};$$

hence the representation (2.32) is unique.

If in addition $\Delta_m > 0$, the sequence b is strictly increasing, and, if $\Delta_i > 0$, $i = 2, \ldots, m-1$, the sequence b_i/b_{i+1} is strictly increasing.

PROOF   To exclude trivialities we may assume $m \geq 3$. Suppose we are given a non-decreasing sequence b with b_i/b_{i+1} non-decreasing. Let us define, $\xi_i := \ln b_i$, $1 \leq i \leq m$. Then, $\xi_1 = 0$ since $b_1 = 1$ and ξ_i satisfies the following constraints,

$$\xi_i \leq \xi_{i+1}, \quad 1 \leq i \leq m-1$$
$$\xi_{i-1} + \xi_{i+1} \leq 2\xi_i, \quad 2 \leq i \leq m-1. \qquad (2.33)$$

We now introduce the new variables,

$$\Delta_i := 2\xi_i - \xi_{i-1} - \xi_{i+1} = \xi_i - \xi_{i-1} - (\xi_{i+1} - \xi_i) \geq 0, \quad 2 \leq i \leq m-1,$$
$$\Delta_m := \xi_2 - \sum_{i=2}^{m-1} \Delta_i = \xi_m - \xi_{m-1} \geq 0, \qquad (2.34)$$

and thus it follows straightforwardly that for $2 \leq i \leq m$,

$$\xi_2 - \sum_{l=2}^{i-1} \Delta_l = \xi_i - \xi_{i-1} \geq 0, \qquad (2.35)$$

with an empty sum defined to be zero. We so obtain by (2.35) for $2 \le i \le m$,

$$\xi_i = \xi_i - \xi_1 = \sum_{k=2}^{i} \xi_k - \xi_{k-1} = \sum_{k=2}^{i} \{\xi_2 - \sum_{l=2}^{k-1} \Delta_l\}$$

$$= (i-1)\xi_2 - \sum_{k=3}^{i} \sum_{l=2}^{k-1} \Delta_l = (i-1)\xi_2 - \sum_{l=2}^{i-1} \sum_{k=l+1}^{i} \Delta_l$$

$$- (i-1)\xi_2 - \sum_{l=2}^{i-1} (i-l)\Delta_l. \tag{2.36}$$

Then via (2.34) and (2.36) we may express ξ_i, and so b_i, in the new parameters Δ_i by

$$\xi_i = (i-1) \sum_{l=2}^{m} \Delta_l - \sum_{l=2}^{i-1} (i-l)\Delta_l = (i-1) \sum_{l=i}^{m} \Delta_l + \sum_{l=2}^{i-1} (l-1)\Delta_l$$

$$= \sum_{l=2}^{m} \min(l-1, i-1)\Delta_l,$$

$$b_i = \exp(\xi_i). \tag{2.37}$$

The converse and uniqueness follow by straightforward derivation of (2.34) for $\xi_i := \ln b_i$ with a sequence (b_i) given by (2.32) for a non-negative sequence $\Delta_2, \ldots, \Delta_m$.

Combining (2.34) and (2.35) yields

$$\xi_i - \xi_{i-1} = \sum_{l=i}^{m} \Delta_l, \qquad 2 \le i \le m.$$

Hence, if b is given by (2.32) with $\Delta_i \ge 0$, the sequence b is over all *strictly* increasing if and only if $\Delta_m = \xi_m - \xi_{m-1} > 0$. Finally, by (2.34) the sequence b_i/b_{i+1} is strictly increasing if and only if $\Delta_i > 0$ for $i = 2, \ldots, m-1$. ◻

COROLLARY 2.4
A correlation structure of type (2.29), generated by a sequence b as in Theorem 2.1, can be represented by

$$\rho_{ij} = \exp\left[-\sum_{l=i+1}^{m} \min(l-i, |j-i|)\Delta_l\right], \qquad 1 \le i, j \le m, \tag{2.38}$$

with $\Delta_i > 0$, $2 \le i \le m$.

REMARK 2.8 For consistency with the notation in (2.29) in connection with a Libor market model, one can take in (2.38), $m = n-1$ and set $\rho_{kl}^{(0)} := \rho_{k+1,l+1}$, $0 \le k, l < n-1$. ◻

As an alternative, one can also derive semi-parametric Libor correlation structures (2.29) from Corollary 2.3 by taking a function β which is, in addition to Corollary 2.3, strictly increasing and such that for each $v \geq 0$ the function $s \rightarrow \beta(s)/\beta(s + v)$ is strictly increasing as well. Indeed, then for any increasing sequence u_k, $0 \leq k < n - 1$, in $\mathbb{R}_+$, the sequence $b_k := \beta(u_k)$, $0 \leq k < n - 1$, defines a Libor correlation structure (2.29). In this context we have the following representation theorem for the correlation generating function β.

THEOREM 2.2
Let $\beta : [0, \infty) \rightarrow \mathbb{R}_+$ be two times continuously differentiable in $(0, \infty)$ with (i): $\beta(0) = \beta(0+) = 1$, (ii): β is non-decreasing, and (iii): for each $v > 0$ the function $s \rightarrow \beta(s)/\beta(s + v)$ is non-decreasing.
  Then, there exists a non-negative continuous function η in $(0, \infty)$, with (iv): $\int_1^\infty \eta(z)dz < \infty$, (v): $\int_{0+}^1 dy \int_y^\infty \eta(z)dz < \infty$, and a constant $C \geq 0$, such that

$$\beta(s) = \exp[\int_0^s dy \int_y^\infty \eta(z)dz + Cs]. \tag{2.39}$$

  Conversely, let $C \geq 0$ and η in $(0, \infty)$ be non-negative, continuous, and such that it holds (iv), (v). Then, (2.39) defines a two times differentiable function in $(0, \infty)$ which satisfies (i), (ii), (iii), and $\beta(0+) = 1$.
  If, in addition, η is strictly positive, then both β and $s \rightarrow \beta(s)/\beta(s+v)$ are strictly increasing for each $v \geq 0$.

PROOF   Let us begin with the converse statement. If C and η are as stated, β is well defined by (2.39). Moreover, by direct verification it follows that *(i)*, *(ii)*, and *(iii)* hold. Indeed, we have

$$\beta'(s) = \beta(s)(\int_s^\infty \eta(z)dz + C) \geq 0, \qquad \text{and} \tag{2.40}$$

$$\begin{aligned}
\frac{d}{du}\frac{\beta(u)}{\beta(u+s)} &= \frac{d}{du}\exp\Big[\int_0^u dy \int_y^\infty \eta(z)dz + Cu \\
&\quad - \int_0^{u+s} dy \int_y^\infty \eta(z)dz - C(u+s)\Big] \\
&= \frac{\beta(u)}{\beta(u+s)}\left(\int_u^\infty \eta(z)dz - \int_{u+s}^\infty \eta(z)dz\right) \\
&= \frac{\beta(u)}{\beta(u+s)}\int_u^{u+s} \eta(z)dz \geq 0.
\end{aligned} \tag{2.41}$$

Now suppose that β is two times continuously differentiable in $(0, \infty)$ and such that *(i)*, *(ii)*, and *(iii)* hold. Then by differentiability of β, we have for any $s, u > 0$,

$$\beta(u + s)\beta'(u) - \beta(u)\beta'(u + s) \geq 0,$$

due to *(iii)*. Hence, for any $s, u > 0$,

$$\frac{\beta'(u)}{\beta(u)} - \frac{\beta'(u + s)}{\beta(u + s)} \geq 0,$$

so β'/β is both non-negative and non-increasing. Hence, we may define $C := \lim_{u \to \infty} \beta'(u)/\beta(u)$ with $C \geq 0$. Since β is two times differentiable in $(0, \infty)$, we can define a function η on $(0, \infty)$ by

$$\eta(u) := -\frac{d}{du}\frac{\beta'(u)}{\beta(u)} = -\frac{d^2}{du^2}\ln\beta(u),$$

which is thus non-negative. It then follows that for $u > 0$,

$$\frac{\beta'(u)}{\beta(u)} = \int_u^\infty \eta(z)dz + C,$$

where in particular the r.h.s. integral exists, hence *(iv)*. Next, by integration and taking into account $\beta(0+) = 1$, we obtain

$$\ln\beta(s) = \int_{0+}^s du \int_u^\infty \eta(z)dz + Cs, \qquad (2.42)$$

and, in particular, existence of the r.h.s. integral, hence *(v)*, and we have (2.39).

Finally, if in addition η is strictly positive, the left-hand sides of (2.40) and (2.41) are strictly positive.

$$\sqcap$$

COROLLARY 2.5

A function β which satisfies the requirements of Theorem 2.2 induces a correlation function (2.10) of the form,

$$\rho(s, u) = \frac{\min[\beta(s), \beta(u)]}{\max[\beta(s), \beta(u)]}$$

$$= \exp\left[-C|u - s| - \Big|\int_s^u dy \int_y^\infty \eta(z)dz\Big|\right].$$

Example 2.1

Let us take $C := 0$ and, for some constants γ, ρ with $0 < \gamma < 1$, $\rho > 0$, the function $\eta(z) := \rho\gamma(1 - \gamma)z^{\gamma - 2}$ in $(0, \infty)$. Then, clearly, *(iv)* and *(v)* of Theorem 2.2 hold and we thus obtain via (2.39) a correlation generating function,

$$\beta(s) := \exp(\rho s^\alpha).$$

So, for instance, the sequence b with $b_i := \beta(i) = \exp(\rho i^\alpha)$, $0 \le i < n-1$, defines a correlation structure (2.29). □

In fact, Theorem 2.2 and Corollary 2.5 are presented because of their nice analogy with Theorem 2.1 and Corollary 2.4. Our further study of correlation structures, however, will be exclusively based on Theorem 2.1 and Corollary 2.4, for the following reason. Theorem 2.1 establishes a one to one correspondence between any correlation generating sequence b in (2.29) and a non-negative sequence Δ, whereas in general there are infinitely many functions β in Theorem 2.2, which may generate the same sequence b.

2.3.4 Generation of Low Parametric Structures

From Corollary 2.4 we can derive conveniently various low parametric structures which all fit into the framework of Libor market models (2.21), equipped with correlation structures of the form (2.29). We give some examples.

Example 2.2
Let us take $\Delta_2 = \ldots = \Delta_{m-1} =: \alpha \ge 0$ and $\Delta_m =: \beta \ge 0$. Then, (2.38) yields the correlation structure

$$\rho_{ij} = \exp\left[-|i-j|(\beta + \alpha(m - \frac{i+j+1}{2}))\right], \quad 1 \le i,j \le m. \quad (2.43)$$

Note that for $\alpha = 0$ we get $\rho_{ij} = e^{-\beta|i-j|}$, a simple correlation structure frequently used in practice in spite of the, in fact, unrealistic consequence that $i \to Cor(\Delta L_i, \Delta L_{i+p})$ is *constant* rather than increasing for fixed p. Let us introduce new parameters $\rho_\infty := \rho_{1m}$ and $\eta := \alpha(m-1)(m-2)/2$, hence

$$\beta = -\frac{\alpha}{2}(m-2) - \frac{\ln \rho_\infty}{m-1} = \frac{-\eta - \ln \rho_\infty}{m-1}.$$

Then (2.43) becomes

$$\rho_{ij} = \exp\left[-\frac{|i-j|}{m-1}(-\ln \rho_\infty + \eta \frac{m-i-j+1}{m-2})\right],$$
$$0 \le \eta \le -\ln \rho_\infty, \quad 1 \le i,j \le m. \quad (2.44)$$

While the structures (2.43) and (2.44) are essentially the same, by the reparametrization of (2.43) into (2.44) the parameter stability is improved: Relatively small movements in the $b-$sequence connected with (2.44), thus the (market) correlations, cause relatively small movements in the parameters ρ_∞

and η. In fact, this can be seen also by analytical comparison of the parameter sensitivities (derivatives) in (2.43) and (2.44). ◻

The following three parametric structure is a refinement of (2.43):

Example 2.3
Suppose $m > 2$ and let Δ_i be a linear function of i for $2 \le i \le m - 1$, with $\Delta_2 =: \alpha_1 \ge 0$, $\Delta_{m-1} =: \alpha_2 \ge 0$. Hence, for $i = 2, \ldots, m-1$, we have

$$\Delta_i = \alpha_1 \frac{m - i - 1}{m - 3} + \alpha_2 \frac{i - 2}{m - 3}$$

and we further define $\Delta_m =: \beta \ge 0$. Then, from (2.38) we get by rather tedious but elementary algebra the correlation structure

$$\rho_{ij} = \exp\left[-|j - i|\left(\beta - \frac{\alpha_2}{6m - 18}(i^2 + j^2 + ij - 6i - 6j - 3m^2 + 15m - 7)\right.\right.$$

$$\left.\left. + \frac{\alpha_1}{6m - 18}(i^2 + j^2 + ij - 3mi - 3mj + 3i + 3j + 3m^2 - 6m + 2)\right)\right]. (2.45)$$

Note that (2.45) collapses to (2.43) for $\alpha_1 = \alpha_2 = \alpha$. As in (2.43) we now re-parameterize (2.45) by $\rho_\infty = \rho_{1m}$ which yields

$$\beta = \frac{-\ln \rho_\infty}{m - 1} - \frac{\alpha_1}{6}(m - 2) - \frac{\alpha_2}{3}(m - 2).$$

In order to gain parameter stability as in (2.44), we set

$$\alpha_1 = \frac{6\eta_1 - 2\eta_2}{(m - 1)(m - 2)}, \quad \alpha_2 = \frac{4\eta_2}{(m - 1)(m - 2)}$$

and then (2.45) becomes

$$\rho_{ij} = \exp\left[-\frac{|j - i|}{m - 1}\left(-\ln \rho_\infty\right.\right.$$

$$+ \eta_1 \frac{i^2 + j^2 + ij - 3mi - 3mj + 3i + 3j + 2m^2 - m - 4}{(m - 2)(m - 3)}$$

$$\left.\left. - \eta_2 \frac{i^2 + j^2 + ij - mi - mj - 3i - 3j + 3m + 2}{(m - 2)(m - 3)}\right)\right], \qquad (2.46)$$

$$1 \le i, j \le m, \quad 3\eta_1 \ge \eta_2 \ge 0, \ 0 \le \eta_1 + \eta_2 \le -\ln \rho_\infty.$$

Obviously, for $\eta_1 = \eta_2 = \eta/2$, (2.46) yields (2.44) again. Correlation structure (2.46) is used for implied calibration procedures in Chapter 3. ◻

Example 2.4
A realistic two-parametric correlation structure. Since calibrating a three parametric structure takes longer than a two-parametric one of course, we

next present an also realistic two parameter structure obtained from (2.46) by taking $\eta_2 = 0$ (equivalently, $\alpha_2 = 0$ in (2.45)),

$$
\begin{aligned}
\rho_{ij} = \exp\Big[&- \frac{|j-i|}{m-1}\Big(-\ln\rho_\infty \\
&+\eta\frac{i^2 + j^2 + ij - 3mi - 3mj + 3i + 3j + 2m^2 - m - 4}{(m-2)(m-3)}\Big)\Big],
\end{aligned}
$$
$$
1 \le i, j \le m, \quad \eta > 0, \ 0 < \eta < -\ln\rho_\infty. \tag{2.47}
$$

For this correlation structure the magnitude of concavity of the sequence $\ln b_i$ in (2.29) is decreasing to zero rather than being constant like in (2.43).   ⧉

Example 2.5

It is easily checked that the sequence b defined by

$$
b_i = e^{\beta(i-1)^\alpha}, \quad 1 \le i \le m, \ \beta > 0, \ 0 < \alpha < 1, \tag{2.48}
$$

satisfies the requirements of Theorem 2.1 and thus yields a correlation structure of the form (2.29), which was proposed in Schoenmakers & Coffey (1998) and used in Kurbanmuradov, Sabelfeld & Schoenmakers (2002) for testing Libor approximations (Chapter 6). By Theorem 2.1, the structure (2.48) has a representation (2.32) with

$$
\begin{aligned}
\Delta_i &= 2\beta(i-1)^\alpha - \beta(i-2)^\alpha - \beta i^\alpha, \quad 2 \le i \le m-1, \tag{2.49} \\
\Delta_m &= \beta(m-1)^\alpha - \beta(m-2)^\alpha,
\end{aligned}
$$

where $\Delta_i > 0$ for $1 \le i \le m$, and Δ_i is strictly decreasing for $2 \le i \le m-1$. By introducing $\rho_\infty := 1/b_m$, we get from (2.48)

$$
\rho_{ij} = e^{\ln\rho_\infty\left|\left(\frac{i-1}{m-1}\right)^\alpha - \left(\frac{j-1}{m-1}\right)^\alpha\right|}, \quad 1 \le i, j \le m, \ 0 < \alpha < 1, \ \beta > 0. \tag{2.50}
$$

In principal, (2.50) has similar properties as the structure (2.47). See further Example 2.1, where basically the same sequence (2.48) is derived from a correlation generating function.   ⧉

REMARK 2.9   Let us now consider the parametrization of market correlations used by Rebonato (1999a), which has for an equidistant tenor structure the following form

$$
\rho_{ij} = \rho_\infty + (1 - \rho_\infty)\exp\left[-|j-i|(\beta - \alpha\max(i,j))\right]. \tag{2.51}
$$

The structure (2.51) can be seen as a perturbation of the correlation structure

$$
\hat{\rho}_{ij} = \rho_\infty + (1 - \rho_\infty)\exp\left[-|j-i|\beta\right] \tag{2.52}
$$

and has the desirable property that

$$i \to \rho_{i,i+p} = \rho_\infty + (1 - \rho_\infty) \exp\left[-p(\beta - \alpha(i + p))\right]$$

is *increasing* for $\alpha > 0$ and thus may produce for a given tenor structure realistic market correlations for properly chosen ρ_∞, $\beta > 0$ and (small) $\alpha > 0$; see Rebonato (1999a). However, while (2.51) may produce realistic correlations, it should be noted that its $(\alpha, \beta, \rho_\infty)$ domain of positivity is not explicitly specified. Hence, for a particular choice of parameters it is not directly guaranteed that (2.51) defines a correlation structure indeed. In particular, it can be verified easily that (2.51) does not fit in the framework of (2.29). It is clear that all the correlation structures consistent with (2.29), in particular (2.43), (2.45), (2.48), and their equivalent representations, do not suffer for this problem as, from the way they are constructed, they are endogenously positive definite and incorporate the economically realistic properties $j \to \rho_{ij}$ is *decreasing* for $j > i$ *and* $i \to \rho_{i,i+p}$ is *increasing* for fixed p, for any choice of parameters *in their well specified domain*. This property makes the correlation structures consistent with (2.29) presented in this section particularly suitable for calibration purposes as search routines are thus prevented for searching in parameter regions where the matrix ρ_{ij} fails to be a correlation structure in fact.   ⬚

2.3.5 Parametric Low Rank Structures

For different reasons, for example computational efficiency, it can be desirable to work with a d-factor Libor model where d is much smaller than n, the number of Libors. For example one wants to use a two or three factor model (d is two or three, respectively). As we are generally interested in parsimonious volatility structures we thus need to generate parsimonious but meaningful parametrisations of low rank correlation structures. As an example, suppose we want to construct a two factor Libor model with volatility structure of type (2.21). Since $d = 2$ we need in (2.21) a $(n-1) \times (n-1)$ correlation matrix $\rho^{(0)}$ of rank two. Now observe that such a correlation matrix can always be written as

$$\rho_{kl}^{(0)} = \left(e_k^{(0)}\right)^\top e_l^{(0)} = \cos(\phi_k - \phi_l), \quad 0 \le k, l < n - 1,$$

$$e_{k1}^{(0)} := \cos(\phi_k) \quad e_{k2}^{(0)} := \sin(\phi_k), \tag{2.53}$$

for some set of angles $\phi_0, \ldots, \phi_{n-2}$ with $\phi_0 := 0$. In fact, the structure (2.53) is a special case ($d = 2$) of the general angle representation for a correlation matrix of rank d. See Rousseeuw and Molenberghs (1993), Rebonato (1999a), Rapisarda, Brigo and Mercurio (2002). The representation (2.53) has still $n - 1$ degrees of freedom and a sensible parametrisation of it comes down to an economically sensible parsimonious parametrisation of the angles ϕ_k. It is clear that such a parametrisation, for instance, in the spirit of (2.29) is not obvious. In Rebonato (1999a) a Monte Carlo based least squares calibration of a certain rank-d angle structure (e.g., (2.53) for $d = 2$) to some fixed target ("market") correlation matrix is proposed. Such a search procedure, however, requires in general relatively many parameters to calibrate, $O(nd)$ in fact. As explained in Chapter 3, for calibration purposes we prefer parsimonious structures and therefore we seek for a suitable way to parametrise a low rank correlation matrix with only a few number of parameters. A natural tool for this is Principal Component Analysis (PCA).

Principal Component Analysis

Let us consider a full rank $(n-1) \times (n-1)$ correlation matrix ρ. Then, for any choice of the factor dimension d, we can map ρ into a rank d correlation matrix by the method of Principal Component Analysis: We take the first d principal components of ρ, i.e., corresponding to the eigenvalues $\lambda_1 > \cdots > \lambda_d \geq 0$ (we assume different eigenvalues), we take unit eigenvectors $u_1, ..., u_d$. Then we define $e_i \in R^d$, $i = 1, ..., n - 1$, by $e_i(p) = u_p(i)\sqrt{\lambda_p}$, $p = 1, ..., d$. Next, in order to get a rank-d *correlation* matrix $\rho^{(d)}$ we re-normalize these vectors and so define $\tilde{e}_i \in R^d$ by $\tilde{e}_i = e_i/|e_i|$, to obtain

$$PCA(\rho; d) := \tilde{\rho}^{(d)} := (\tilde{e}_i^\top \tilde{e}_j), \quad 1 \leq i, j < n.$$

Let us start with a full rank parametric correlation structure, for example we take the structure $\rho_{ij}^{\eta_1, \eta_2, \rho_\infty}$ in (2.46). Then, by PCA one can produce for any choice of d the correlation structure

$$\tilde{\rho}_{ij}^{(\eta_1, \eta_2, \rho_\infty; d)} := PCA(\rho^{\eta_1, \eta_2, \rho_\infty}, d), \tag{2.54}$$

being a rank-d correlation structure parametrized by three parameters η_1, η_2, ρ_∞.

REMARK 2.10   It should be noted that the rank-d correlation structure, resulting from the operation

$$\rho^{\eta_1, \eta_2, \rho_\infty} \to PCA(\rho^{\eta_1, \eta_2, \rho_\infty}, d),$$

usually cannot be given by some relatively simple analytic expression such as (2.46) which we have for $\rho^{\eta_1, \eta_2, \rho_\infty}$. However, within a certain calibration procedure involving the parameters $\eta_1, \eta_2, \rho_\infty$ for instance, the PCA operation can be carried out effectively by standard numerical procedures; see for example, *Numerical Recipes,* Press et al. (1992).   ⬚

2.4   Some Possible Applications of Parametric Structures

2.4.1   Smoothing Historical Libor Correlations (sketch)

We assume a fixed time tenor structure $T_0 < T_1 < T_2 < ... < T_n$ with equi-distant tenors, i.e., $\delta = T_{i+1} - T_i$ for all i, and a system of risk-free zero-coupon bonds B_i also called "present values" which mature at T_i, hence $B_i(T_i) - 1$. We further suppose that up to the present calendar date T_0 time series of historical data of the present values $B_i(t)$, $t \leq T_0$, $i = 1, \ldots, n$ are available, for instance on a daily basis. These values can be obtained from the present value curve $T \to B(t, T)$, $t \leq T_0$, $T \geq T_0$, which in turn may be constructed by interpolating the available present values at each historical date $t \leq T_0$. For more details on yield curve smoothing methods, see, e.g., Fisher et al. (1995), Linton et al. (1998). So, in principle, time series of forward Libor rates are available via

$$L_i(t) = (\frac{B_i(t)}{B_{i+1}(t)} - 1)\delta^{-1}, \quad t \leq T_0, \ i = 1, \ldots, n - 1.$$

As a rough approximation of the Libor market model, by neglecting the drifts in fact, we assume that returns of the forward Libor rates may be modelled by a multi-dimensional Gaussian distribution. We thus assume

$$\frac{\Delta L_i(t)}{L_i(t)} \approx \gamma_i^\top \Delta Z \sqrt{\Delta t},$$

where $\gamma_i \in \mathbb{R}^{n-1}$ is assumed to be constant over the time interval where the time series are considered, and ΔZ is an $(n - 1)$-dimensional standard Gaussian vector, i.e., $\mathbb{E}\,\Delta Z_i \Delta Z_j = \delta_{ij}$. We so have

$$\mathbb{E}\, \frac{\Delta L_i}{L_i} \frac{\Delta L_j}{L_j} = \gamma_i^\top \gamma_j \Delta t = |\gamma_i||\gamma_j|\rho_{ij}\Delta t. \tag{2.55}$$

In (2.55) the $|\gamma_i|$ can be identified by historical estimates of the left-hand side for $i = j$, since $\rho_{ii} = 1$. Then, in (2.55) we plug in the estimations for $|\gamma_i|$ and some parametric correlation structure, for example one of the full rank structures (2.46), (2.47), or some parametric low rank structure (2.54) obtained via Principal Component Analysis. Finally, we calibrate via least squares the parameters of this correlation structure to the left-hand side estimations of (2.55) for $i \neq j$. Resuming we propose the following program for identification of a smooth Libor return correlation structure:

I *Construct daily time series for the different forward Libors*

II *Estimate historically the left-hand side of (2.55) in the usual way and thus obtain for $i = j$ in particular historical estimations for the $|\gamma_i|$.*

III *Fit the (for instance two or three) parameters of a parametric correlation structure ρ via (2.55) by standard least squares to the historical
estimations obtained for $i \neq j$.*

2.4.2   Calibration to Caps and Swaptions by Using Historical Correlations (sketch)

One can take in Section 2.4.1 the historically identified instantaneous correlation structure as fixed $\rho^{(0)}$ in the volatility structure (2.21) of a Libor
market model and then calibrate some parametric scalar volatility function
g to the prices of swaptions via a least squares objective function proposed
in Section 3.2.1. For instance, one could take the parametric form (2.25).
Finally, after identifying g the coefficients c_i can be obtained by fitting the
caplets via (2.19).

We note, however, that this calibration method involves historical estimation of Libor correlations which, on the one hand, may be a rather elaborate
job and, on the other hand, is not really desirable in the view of implied
calibration enthusiasts who follow the idea that the Libor model should be
identified by exclusively using price information relevant at the calibration
date (T_0). Full implied calibration methods are described next in Chapter 3.

Chapter 3

Implied Calibration of a Libor Market Model to Caps and Swaptions

3.1 Orientation and General Aspects

The subject of this chapter is robust calibration of the Libor Market Model (2.1), one of the main themes in this book. Our goal is in particular stable implied calibration to caps and swaptions without using an explicitly given (e.g., historical) input correlation matrix. We first consider some general aspects.

Let us suppose that the scalar volatilities σ_i and correlations ρ_{ij} corresponding to (2.1) are piece-wise constant functions of time. Clearly, then the whole system of scalar volatilities $\{\sigma_i : 1 \leq i < n\}$ contains $n(n-1)/2$ constants to be determined. On the other hand the cardinality of a complete set of market caplet and swaption volatility quotes, denoted by $\mathcal{M} := \{\widehat{\sigma}_{p,q}^B : 1 \leq p < q \leq n\}$ with $\widehat{\sigma}_p^B := \widehat{\sigma}_{p,p+1}^B$ being the caplet volatilities, is equal to $n(n-1)/2$ also. So we see immediately that in general the model (2.1) is over-determined. Indeed, one may prespecify *any* set of piece-wise constant correlation functions (i.e., a correlation structure), and then there are still enough degrees of freedom left for matching the set $\mathcal{M}$. So for an unambiguous model identification we need to reduce in (2.1) the high degree of freedom. However, this reduction should be economically motivated of course.

As a first attempt let us take σ_i and ρ_{ij} to be time independent. We so have $n-1$ scalar volatility parameters and $(n-1)(n-2)/2$ correlation parameters, hence a total of $n(n-1)/2$ which is equal to $\#\mathcal{M}$. However, despite the fact that identification of this $n(n-1)/2$ parameter model may cause serious stability problems, a Libor model with time independent scalar volatilities σ_i is considered unrealistic as in practice Libor volatilities tend to increase shortly before they approach their maturity.

An economically more relevant structural assumption on the piece-wise volatility structure is the requirement of time shift homogeneity in some sense. For instance, let us consider the Hull-White volatility structure (2.14). This structure requires the determination of $n-1$ constants $\Lambda_i, 0 \leq i < n-1$ and

$(n-1)(n-2)/2$ correlations $\rho_{ij}^{(0)}$, $0 \leq i, j < n-1$. While the homogeneity in the Hull-White structure is quite appealing, the structure has some serious drawbacks. For instance, exact caplet fitting, something where the Libor market model is made for in a sense, usually breaks down due to the fact that condition (2.17) is violated in practice. For this reason we proposed in Section 2.2.5 a quasi-homogeneous volatility structure (2.21).

A natural calibration procedure for (2.21) is as follows. First calibrate g and the coefficients c_i in (2.21) to the system of caplets via (2.20) as described in Section 2.2.5. By this procedure scalar volatilities are forced to be "as time homogeneous as possible". Then keep the thus identified scalar volatility functions fixed and calibrate a suitably parametrised correlation structure $\rho^{(0)}$ to swaption prices. Although this procedure does not suffer for principle stability problems essentially, the attainable fit is not always very good in practice; see for details Section 3.5. On the other hand, as we expose in Section 3.2.1, a straightforward (e.g., least squares) calibration of (2.21) to the set $\mathcal{M}$ of caplets and swaptions, which involves joint identification of the scalar volatility function g and correlation structure $\rho^{(0)}$, is unstable. We will show that this instability is of intrinsic nature and emerges even when g and $\rho^{(0)}$ are parsimoniously parametrized. Therefore, as an alternative approach, instead of imposing (almost) time homogeneity of scalar volatilities, we propose in Section 3.4 a regularisation of the least squares calibration routine via a "Market Swaption Formula", a rule-of-thumb formula which reflects from a practitioner's point of view an intuitive relationship between caplet volatilities and swap volatilities. By several case studies with practical data we show that with a thus regularised least squares procedure, calibration of the structure (2.21) with a structure suitably parametrized function g and correlation structure $\rho^{(0)}$ to caps and swaptions can be kept stable while attaining a satisfactory calibration fit. The involved calibration routines are based on standard swaption approximations or their refinements given in Section 1.3.3.

As already discussed in Section 1.3.4, we restrict ourselves to a standard Libor market model and so it will be not possible to model volatility skews, hence to calibrate the model to a complete cap-strike matrix. Instead, if a particular exotic product is to be evaluated, we suggest to choose for the calibration in the cap(let)-strike matrix for each maturity a quote with a certain strike, such that the family of selected cap(let)s with the respective strikes is in some sense well related to the exotic product under consideration. By default we usually take ATM cap(let)s as calibration instruments.

3.2   Assessment of the Calibration Problem

As suggested in several studies, e.g., Schoenmakers & Coffey (1998), Rebonato (2000), instead of calibrating a Libor Market Model (2.1) with volatility structure (2.21) directly to market prices via Monte Carlo simulation for instance, we may equivalently calibrate it such that the system of implied Black volatilities of model caplet/swaption prices is consistent with the system of Black volatilities due to the corresponding market prices.[1] Further we suppose that the simple or refined swaption approximation formula, (1.41) or (1.42) respectively, is good enough in this context. We thus seek for a scalar volatility function g, coefficients c_i, $1 \leq i < n$, and an initial correlation matrix $\rho^{(0)}$ such that the following relationship with the market Black volatilities of swaptions, denoted by $\widehat{\sigma}_{p,q}^B$, holds at least in a good approximation,

$$(\widehat{\sigma}_{p,q}^B)^2 \doteq \sum_{i,j=p}^{q-1} \frac{w_i^{p,q}(0)w_j^{p,q}(0)L_i(0)L_j(0)}{S_{p,q}^2(0)} .$$

$$\cdot \frac{c_i c_j}{T_p} \int_0^{T_p} g(T_i - t)g(T_j - t)\rho_{i-m(t),j-m(t)}^{(0)} dt,$$

$$1 < p < q \leq n, \tag{3.1}$$

with $w_i^{p,q}(0)$ as defined in (1.41) for the simple swaption approximation or as in (1.42) for the refined approximation. Note that for $1 \leq p < n$, $w_p^{p,p+1} = 1$ and, since one period swaptions are caplets in fact, $\widehat{\sigma}_p^B := \widehat{\sigma}_{p,p+1}^B$ are market caplet volatilities for which (3.1) yields

$$(\widehat{\sigma}_p^B)^2 \doteq \frac{c_p^2}{T_p} \int_0^{T_p} g^2(T_p - t)dt, \quad 1 \leq p < n. \tag{3.2}$$

We thus can solve the c_i's explicitly from g and the market caplet volatilities by (3.2). So by introducing for $p \leq \min(i,j)$ the quantities

$$\alpha_{i,j,p}^{g,\rho^{(0)}} := \frac{\sqrt{T_i}\sqrt{T_j} \int_0^{T_p} g(T_i - t)g(T_j - t)\rho_{i-m(t),j-m(t)}^{(0)} dt}{T_p \sqrt{\int_0^{T_i} g^2(s)ds}\sqrt{\int_0^{T_j} g^2(s)ds}}, \tag{3.3}$$

we can re-express (3.1) as

$$(\widehat{\sigma}_{p,q}^B)^2 \doteq \sum_{i,j=p}^{q-1} \frac{w_i^{p,q}(0)w_j^{p,q}(0)L_i(0)L_j(0)}{S_{p,q}^2(0)} \widehat{\sigma}_i^B \widehat{\sigma}_j^B \alpha_{i,j,p}^{g,\rho^{(0)}},$$

$$1 \leq p < q \leq n. \tag{3.4}$$

---

[1] Since we calibrate to at the money (ATM) caps and swaptions this makes hardly any difference in practice.

Note that $\alpha_{p,p,p}^{g,\rho^{(0)}} \equiv 1$ and so (3.4) is trivially fulfilled for $q = p + 1$.

3.2.1 Straightforward Least Squares, Stability Problems

As a first and canonical approach for calibrating g and $\rho^{(0)}$ via (3.4) we discuss the method of (straightforward) least squares calibration. With

$$\sigma_{p,q}^B(g, \rho^{(0)}) := \sqrt{\sum_{i,j=p}^{q-1} \frac{w_i^{p,q}(0)w_j^{p,q}(0)L_i(0)L_j(0)}{S_{p,q}^2(0)} \widehat{\sigma}_i^B \widehat{\sigma}_j^B \alpha_{i,j,p}^{g,\rho^{(0)}}} \qquad (3.5)$$

being, at least in a very good approximation, the implied swaption volatilities due to Libor model (2.21) with volatility function g and initial correlation matrix $\rho^{(0)}$, we consider minimization of the following "root mean squares" distance,

$$RMS(g, \rho^{(0)}) :=$$

$$\sqrt{\frac{2}{(n-1)(n-2)} \sum_{1 \le p \le q-2, \ q \le n} \left(\frac{\widehat{\sigma}_{p,q}^B - \sigma_{p,q}^B(g, \rho^{(0)})}{\widehat{\sigma}_{p,q}^B} \right)^2} \longrightarrow \min_{g, \rho^{(0)}} . \qquad (3.6)$$

Loosely speaking, via the minimization procedure (3.6) one tries to identify g and $\rho^{(0)}$ such that the model swaption volatility surface $\{\sigma_{p,q}^B(g, \rho^{(0)}) : 1 \le p < q \le n\}$ is as good as possible in accordance with the market swaption volatility surface $\{\widehat{\sigma}_{p,q}^B : 1 \le p < q \le n\}$. Now we observe that the market volatility surface intersects with the model volatility surface at the intersection line $\{\widehat{\sigma}_{p,p+1}^B : 1 \le p < n\}$ *for any choice of g and $\rho^{(0)}$.* Further, in practice typical market swaption volatility surfaces are rather flat (see Figure 3.1(a)) and, for not too erratic functions g and correlation structures $\rho^{(0)}$, the corresponding model volatility surfaces are rather flat as well; see Figure 3.1(b) for a model volatility surface fitted to (a). As a consequence, the main effect of varying both g and $\rho^{(0)}$ is just a rotation of the model surface around the intersection "line" $\{\widehat{\sigma}_{p,p+1}^B : 1 \le p < n\}$. For an illustration see Figure 3.2(a) and Figure 3.2(b), where we have taken g and $\rho^{(0)}$ to be parametric structures of the form (2.25) and (2.46), respectively. So, basically, it is possible to rotate the model swaption volatility surface to the market volatility surface by the tuning of two main influential objects g and $\rho^{(0)}$. Therefore, one could roughly rotate the model surface to the market surface, by choosing two completely different pairs, say $(g, \rho^{(0)})$ and $(\widetilde{g}, \widetilde{\rho}^{(0)})$. Next, by perturbing $(g, \rho^{(0)})$ and $(\widetilde{g}, \widetilde{\rho}^{(0)})$, respectively, both calibrations could be improved to a satisfactory level. *Hence the above calibration procedure is intrinsically instable.* In Section 3.2.2 and Section 3.3.2 we provide theoretical and practical evidence of this instability.

TABLE 3.1:   Lab caplet volatilities (%)

Mat.(yr)	Cplt. Vols.	Mat.(yr)	Cplt. Vols.
0.5	14.62	10.5	11.42
1	14.01	11	11.41
1.5	13.34	11.5	11.39
2	12.86	12	11.38
2.5	12.54	12.5	11.36
3	12.30	13	11.35
3.5	12.13	13.5	11.34
4	12.00	14	11.33
4.5	11.90	14.5	11.32
5	11.82	15	11.31
5.5	11.75	15.5	11.30
6	11.69	16	11.30
6.5	11.65	16.5	11.29
7	11.60	17	11.28
7.5	11.57	17.5	11.28
8	11.54	18	11.27
8.5	11.51	18.5	11.26
9	11.48	19	11.26
9.5	11.46	19.5	11.25
10	11.44	20	11.25

3.2.2   Stability Problems in the Laboratory

We are going to show that the stability problems around straightforward least squares calibration as mentioned in Section 3.2.1 are really of intrinsic nature. For this we consider a fixed "laboratory" Libor model (2.1) with practically representative initial yield curve and input volatility structure of type (2.21) and then try to recover the input parameters from least squares calibration to a family of cap and swaption prices, computed within this model. In particular, we consider a forty period semi-annual model, i.e., $n = 41, \delta_i \equiv 0.5, i = 0, \ldots, 40$, with initial quarterly Libor curve taken to be flat 6.091% (corresponding to flat 6.000% under continuously compounding). Further, we take in (2.21) a scalar volatility function g according to (2.25) with $a = 2, b = 3, g_\infty = 0.85$, an initial correlation structure $\rho^{(0)}$ as in (2.46) with $\eta_1 = 1.5, \eta_2 = 0, \rho_\infty = 0.2$, and constants $c_i \equiv 0.13$. For the thus specified Libor model we compute the Black caplet volatilities directly by (3.2) and a set of (ATM) swaption volatilities by Monte Carlo simulation. The results are given in Table 3.1 and Table 3.2, respectively. In view of the (in-)accuracy of the swaption approximation formula used in our calibration procedure later on, the swaptions are simulated with an accuracy of about 0.3% relative based on one standard deviation. This accuracy is attained by simulating 250000 samples for each swaption.

As a first experiment we perform a standard least squares calibration to

TABLE 3.2:   Lab swaption volatilities (%) with noise 0.3% RMS

Mat.(yr)	Per. 1 yr	Per. (yr)	Mat. 1 yr	Mat.(yr)	End 16 yr
1	12.82	1	12.82	1	8.95
2	12.05	2	11.54	2	8.96
3	11.67	3	10.86	3	9.10
4	11.40	4	10.34	4	9.21
5	11.31	5	10.02	5	9.38
6	11.27	6	9.76	6	9.55
7	11.21	7	9.61	7	9.63
8	11.23	8	9.40	8	9.84
9	11.16	9	9.30	9	10.00
10	11.19	10	9.21	10	10.09
11	11.08	11	9.11	11	10.26
12	11.09	12	9.04	12	10.48
13	11.07	13	8.98	13	10.64
14	11.14	14	8.95	14	10.76
15	11.09	15	8.95	15	11.09

TABLE 3.3:   LSq fit to lab model

Pars.	Mod. Val.	LSQ Calib.
η_1	1.50	1.44
η_2	0.00	0.21
ρ_∞	0.20	0.19
a	2.00	1.62
b	3.00	3.46
g_∞	0.85	0.74
RMS	0.4%	0.3%

the system of model caplets and swaptions given in Table 3.1 and Table 3.2 via (3.6). To this end we carry out a global search according to Algorithm 3.1 and find a 0.3% RMS parameter fit given in Table 3.3, 3rd column. Apparently, the model parameters are roughly retrieved (with an exception for the correlation parameter η_2). In the 2nd column, we see that the model parameters itself give rise to an RMS error of 0.4%, so a bit larger than the calibration error of 0.3%. This is not surprising since the RMS error (3.6) due to the model parameters involves *both* the noise on the swaption volatilities *and* the error due to the swaption approximation formula (3.1). In fact, in the 0.3% calibration the inaccuracy of the swaption approximation has not been taken into account. Therefore, if we would re-simulate model swaption volatilities due to the parameters in column 3 with an extremely high Monte Carlo precision, we may expect that the relative RMS distance between these re-simulated volatilities and the volatilities in Table 3.2 will be around 0.4% also.

ALGORITHM 3.1

Global search procedure

(i) *Through a number of 6-dimensional Halton points* $(a, b, g_\infty, \eta_1, \eta_2, \rho_\infty)$, *for instance 3000 points, in a representative 6-parameter region;*

(ii) *Consider a subset of "good candidates", for instance the 10 best points due to the 10 lowest values of the objective function;*

(iii) *Start from each point in (ii) a local search (e.g., Powel or Levenberg-Marquardt);*

(iv) *Take the minimum of the local minima in (iii) as result.*

In the next experiment we add some more noise on the swaption volatilities of the lab model by re-simulating these volatilities with a lower number of Monte Carlo samples in order to study the stability of the calibration. By using 10000 samples for each swaption we get a lower RMS accuracy of 1.6%. The noisy volatilities are given in Table 3.4. In the same way as in the previous experiment we now carry out a standard least squares calibration to caplet volatilities and noisy swaption volatilities given in Table 3.1 and Table 3.4, respectively. The results are given in Table 3.5, column 3. Again we see that the input parameters are roughly retrieved with an RMS error close to the error of 1.7% due to the actual model parameters. However, by restricting the search routine to the region $\rho_\infty > 0.5$ we obtain an RMS fit of 1.9% (column 4), so only a little bit larger than the 1.7% due to the actual model parameters. If we imagine that these noisy swaption volatilities would be market quotes in fact, it would be hard to justify on the basis of these nearby fitting errors that the parameters in column 3 should be preferred to the parameters in column 4. So we conclude that the lab model with noise can be fitted with completely different parameter combinations up to a fitting error comparable with the noise magnitude. On the other hand we observe that the least squares search routine is still able to recover the input parameters more or less, but this is due to the fact that we have a lab model in the sense that swaption volatilities are consistent with a model for a particular set of model parameters, disturbed with independent noise. This is explained as follows. For simplicity, we disregard the inaccuracy of the swaption approximation formula. If v_0 is the set of model input parameters yielding model swaption volatilities $\sigma_i(v_0)$, and $\hat{\sigma}_i := \sigma_i(v_0) + \epsilon_i$ are noisy volatilities with zero mean i.i.d. noise ϵ_i, $i = 1, \ldots, K$, with K being the number of swaptions, then

$$E \sum_{i=1}^{K} (\hat{\sigma}_i - \sigma_i(v))^2 = \sum_{i=1}^{K} (\sigma_i(v_0) - \sigma_i(v))^2 + \sum_{i=1}^{K} E\epsilon_i^2, \tag{3.7}$$

TABLE 3.4:   Lab swaption volatilities (%) with noise 1.6% RMS

Mat.(yr)	Per. 1 yr	Per. (yr)	Mat. 1 yr	Mat.(yr)	End 16 yr
1	13.06	1	13.06	1	8.69
2	11.94	2	11.42	2	9.08
3	12.02	3	11.19	3	9.38
4	11.12	4	10.26	4	9.21
5	11.08	5	10.00	5	9.47
6	11.10	6	9.55	6	9.70
7	11.08	7	9.54	7	9.62
8	11.08	8	9.60	8	9.90
9	10.85	9	9.28	9	9.93
10	11.09	10	9.20	10	9.95
11	11.25	11	8.88	11	10.14
12	11.18	12	8.88	12	10.56
13	11.13	13	8.97	13	10.84
14	11.12	14	9.05	14	10.84
15	10.88	15	8.69	15	10.88

which has a global minimum for $v = v_0$. When dealing with real quoted swaption volatilities in practice however, there will be generally no parameter combination v_0 such that $\hat{\sigma}_i - \sigma_i(v_0)$ can be considered as zero mean i.i.d. noise and then, as a consequence, the stability problems become more serious as we will demonstrate in Section 3.3.2. We conclude this subsection with an important remark.

REMARK 3.1   Due to the fact that we carried out the calibration experiments for a Lab model, with endogenously consistent cap(let), and swaption prices, it is clear that the stability problems encountered by LSq calibration have nothing to do with possible inconsistency between cap(let) and swaption prices. Putting it differently, the same stability problems will be encountered when we consider a Libor model with forward periods consistent with swaption settlement periods, where caplets coincide with one period swaptions, and then calibrate to swaptions only. Hence, in practice, stability problems occurring in a calibration may not be directly blamed on possible misalignments between caps and swaptions!   ∏

3.3   LSq Calibration and Stability Issues in Practice

In this section we outline the practical implementation of the least squares calibration procedure (LSq) and investigate the stability issues discussed in

TABLE 3.5:   LSq fit to noisy lab model

Pars.	Mod. Val.	LSQ Calib.	LSQ, $\rho_\infty > 0.5$
η_1	1.50	1.44	0.59
η_2	0.00	0.21	0.03
ρ_∞	0.20	0.19	0.54
a	2.00	1.49	0.05
b	3.00	3.20	0.83
g_∞	0.85	0.74	0.46
RMS	1.7%	1.6%	1.9%

Section 3.2 for different case studies.

In Section 3.3.1 we deal with practical issues which have to be resolved for implementation of the LSq procedure based on given market data. In particular, we show for typical sets of Euro market data how yield curve data and cap volatility quotes, as given in their usual incomplete form, can be interpolated and converted into a complete system of initial Libor rates and caplet volatilities on a suitable tenor structure. This is necessary for implementation of a LSq procedure as in Section 3.2. In this respect we underline that any Libor model based on a thus constructed system of initial Libor rates and caplet volatilities will be perfectly in consistence with the given yield curve data and cap volatilities. In the calibration to swaptions we will use exclusively quoted swaption volatilities, however.

The identification problem of the Libor model, addressed in Section 3.2.1 and encountered in Section 3.2.2 in connection with an artificial lab model, is considerably amplified when we consider real market data. We will demonstrate this fact in Section 3.3.2 by the data sets given in Section 3.3.1.

3.3.1   Transformation of Market Data and LSq Implementation

We consider four given sets of market data at the dates 14.05.02, 03.06.02, 01.07.02, 08.08.02, respectively. For each date we are given a set of continuously compounded zero rates, a set of ATM cap volatilities, and a swaption volatility matrix. See Tables 3.6, 3.7, 3.8. Due to the given data, the Libor model we are going to calibrate for each respective date will be considered at a semi-annual tenor structure, $T_1 = 0.5\text{yr.}$, $T_2 = 1\text{yr.}$, ..., $T_{40} = 20\text{yr.}$ (so $n = 40$ here), seen from the initial calibration date identified with $T_0 := 0$. The identification of a Libor model on this tenor structure, according to the procedure described in Section 3.2, requires an initial Libor curve and a sequence of caplet volatilities. The initial Libor curve will be computed via definition (1.12) from the present value (zero bond) curve $B(0, T)$, which in turn follows from the continuously compounded zero yield curve $R(0, T)$ by

$$B(0, T) = \exp(-TR(0, T)),$$

where T ranges over $\{T_1, \ldots, T_{40}\}$. The yield curve is for each date 14.05.02, 03.06.02, 01.07.02, 08.08.02, $R(t, T_i)$, $i = 1, \ldots, 40$, obtained from Table 3.6, where the missing values are computed by cubic splines and the so resulting Libor curves are given in Table 3.9.

Whereas in the Lab model in Section 3.2.2 caplet volatilities were directly available via (3.2), we now have to extract caplet volatilities from the given (ATM) cap volatilities. This is done by a procedure called "caplet stripping" which is described below.

Stripping Caplet Volatilities from ATM Caps

Consider at a certain calibration date for each $k = 2, \ldots, n$ the ATM cap over period $[T_1, T_k]$ with price $C_{1,k}^{ATM}$. By convention, the ATM strike rate of each cap is equal to the swap rate $S_{1,k}$ over the accrual cap period. Usually the ATM cap prices are implicitly given by a sequence of "all-in" cap volatilities $\sigma_{1,k}^{ATM}$ such that cap premia are computed by

$$C_{1,k}^{ATM} = \sum_{j=1}^{k-1} C_j(L_j, S_{1,k}, \sigma_{1,k}^{ATM}, T_j), \qquad (3.8)$$

for $k = 2, .., n$. In (3.8), L is the initial Libor curve and C_j denotes the price of caplet j due to strike $S_{1,k}$ and "caplet volatility" $\sigma_{1,k}^{ATM}$ given by Black's formula

$$C_j(L_j, S_{1,k}, \sigma_{1,k}^{ATM}, T_j) := B_{j+1}(\delta_j L_j \mathcal{N}(d_+) - \delta_j S_{1,k} \mathcal{N}(d_-)) \qquad (3.9)$$

with $\mathcal{N}$ being the standard normal distribution and

$$d_\pm = \frac{\ln(L_j/S_{1,k}) \pm \frac{1}{2}(\sigma_{1,k}^{ATM})^2 T_j}{\sigma_{1,k}^{ATM} \sqrt{T_j}}.$$

Note that $\sigma_{1,k}^{ATM}$ is generally not equal to the actual j-th caplet volatility, which is the scalar Libor volatility σ_j in fact. Hence the terms in (3.9) are generally inconsistent with actual caplet premiums. Therefore, the implied volatility $\sigma_{1,k}^{ATM}$ is merely considered as an "all-in" volatility such that (3.8) yields the ATM cap value $C_{1,k}$.

We now want to find a sequence of caplet volatilities σ_j^{Capl}, $j = 1, \ldots, n-1$, such that

$$C_{1,k}^{ATM} = \sum_{j=1}^{k-1} C_j(L_j, S_{1,k}, \sigma_j^{Capl}, T_j) \qquad (3.10)$$

for $k = 2, .., n$. In principal we may solve for these volatilities in an iterative way. Firstly, σ_1^{Capl} is solved from

$$C_{1,2}^{ATM} = C_1(L_1, S_{1,2}, \sigma_{1,2}^{ATM}, T_1) = C_1(L_1, L_1, \sigma_1^{Capl}, T_1).$$

Hence, since $S_{1,2} = L_1$, we have $\sigma_1^{Capl} = \sigma_{1,2}^{ATM}$. Let us suppose that for some k with $k < n$, the volatilities σ_j^{Capl}, $j = 1, .., k-1$ are already obtained. Then σ_k^{Capl} can be solved from the equation

$$C_k(L_k, S_{1,k+1}, \sigma_k^{Capl}, T_k) = C_{1,k+1}^{ATM} - \sum_{j=1}^{k-1} C_j(L_j, S_{1,k+1}, \sigma_j^{Capl}, T_j). \quad (3.11)$$

REMARK 3.2   Usually, there is also given a cap-strike matrix and a similar stripping procedure applies for extracting caplet volatilities σ_i^K consistent with a column of caps in this matrix due to a certain strike K. This procedure is natural for calibrating to strike K cap and swaption volatilities (e.g., ITM or OTM). The description of the strike K stripping procedure and the according calibration method is completely clear from the description in the ATM cases and therefore omitted. ▯

Since the identification of a Libor model according to Section 3.2 requires a complete system of caplet volatilities on the tenor structure, we need also a complete system of ATM cap volatilities. Therefore we interpolate the cap volatilities in Table 3.7 in a smooth way by cubic splines. In this respect a smooth interpolation method is desirable because the stripping procedure based on (3.11) acts like a derivative in a sense. The caplet stripping method applied to the thus interpolated cap volatilities yields for each calibration date a family of caplet volatilities which is listed in Table 3.10.

As a next practical issue we have to take into account the fact that in Euro markets caps are semi-annual whereas swaps are settled annually. Therefore, we shall apply the refined swaption approximation procedure for this case, given in (1.42), which leads to the following LSq objective function.

$$RMS(g, \rho^{(0)}) := \sqrt{\frac{1}{\#\mathcal{S}} \sum_{(p,q)\in\mathcal{S}} \left(\frac{\widehat{\sigma}_{p,q}^B - \sigma_{p,q}^B(g, \rho^{(0)})}{\widehat{\sigma}_{p,q}^B} \right)^2} \longrightarrow \min_{g,\rho^{(0)}}, \quad (3.12)$$

where $\mathcal{S}$ denotes the set of market swaption volatilities $\widehat{\sigma}_{p,q}^B$ we want to calibrate to, and $\sigma_{p,q}^B(g, \rho^{(0)})$ denotes the refined approximated swaption volatility of swaption $(p, q) \in \mathcal{S}$ due to a Libor model with input volatility structure determined by g and $\rho^{(0)}$.

3.3.2   LSq Calibration Studies, Stability Problems in Practice

For the sets of Euro market data in Section 3.3.1 we will now investigate the accuracy and stability of LSq calibration to caps and swaptions. Again

TABLE 3.6:   Cont. comp. spot zero-rates (%)

Mat.(yr)	14.05.02	03.06.02	01.07.02	08.08.02
0.5	3.657	3.638	3.529	3.331
1	4.017	3.987	3.729	3.351
1.5	4.319	4.264	3.972	3.532
2	4.545	4.486	4.145	3.653
2.5	4.676	4.609	4.281	3.812
3	4.789	4.725	4.399	3.953
3.5	4.889	4.812	4.518	4.082
4	4.976	4.894	4.624	4.200
4.5	5.064	4.971	4.702	4.304
5	5.141	5.045	4.771	4.398
5.5	5.208	5.112	4.847	4.488
6	5.269	5.179	4.918	4.572
6.5	5.317	5.232	4.985	4.646
7	5.357	5.282	5.043	4.713
7.5	5.398	5.325	5.102	4.779
8	5.433	5.366	5.154	4.840
8.5	5.477	5.401	5.198	4.879
9	5.514	5.435	5.237	4.913
9.5	5.536	5.464	5.271	4.954
10	5.555	5.492	5.302	4.991
12	5.658	5.583	5.402	5.107
15	5.751	5.699	5.491	5.229
20	5.839	5.802	5.557	5.341
25	5.852	5.807	5.535	5.339

TABLE 3.7:   ATM cap vols. (%), cap start 0.5 yr.

Cap End (yr)	14.05.02	03.06.02	01.07.02	08.08.02
1	17.8	18.5	18.7	23.5
2	17.7	18.3	18.9	23.7
3	17.3	17.5	18.7	21.5
4	16.5	16.8	18	20
5	15.9	16.1	17.3	18.7
6	15.5	15.6	16.6	17.8
7	15.1	15.2	16.2	17
8	14.7	14.9	15.8	16.4
9	14.4	14.6	15.4	15.9
10	14.2	14.3	15.1	15.5
12	13.7	13.8	14.4	14.9
15	13.2	13.3	14	14.2
20	12.5	12.7	13.3	13.4

TABLE 3.8: ATM swaption vols. (%)

Per. (yr) → / Mat. ↓	1	2	3	4	5	6	7	8	9	10	15
				14	**05**	**02**					
1	17.2	15.8	14.6	13.8	13.0	12.6	12.2	11.8	11.5	11.2	10.6
2	15.6	14.3	13.4	12.8	12.3	12.0	11.7	11.4	11.1	10.9	10.3
3	14.6	13.3	12.6	12.1	11.8	11.5	11.3	11.1	10.9	10.7	10.1
4	13.8	12.6	12.1	11.7	11.4	11.2	11.1	10.8	10.6	10.5	9.9
5	12.9	12.2	11.7	11.3	11.0	10.9	10.7	10.6	10.4	10.3	9.8
7	12.1	11.4	11.0	10.7	10.5	10.3	10.2	10.0	9.9	9.8	
10	11.2	10.6	10.3	10.0	9.8	9.7	9.6	9.5	9.5	9.4	
15	9.7	9.4	9.3	9.2	9.1						
				03	**06**	**02**					
1	17.4	16.1	14.9	13.9	13.2	12.7	12.3	11.9	11.6	11.3	10.6
2	15.8	14.5	13.7	12.9	12.5	12.2	11.8	11.5	11.3	11.0	10.4
3	14.9	13.7	12.9	12.3	12.0	11.7	11.5	11.3	11.0	10.8	10.2
4	14.1	12.9	12.3	11.9	11.6	11.4	11.2	11.0	10.8	10.6	10.0
5	13.2	12.4	11.9	11.4	11.2	11.0	10.9	10.7	10.5	10.4	9.8
7	12.2	11.7	11.2	10.8	10.5	10.4	10.3	10.1	10.0	9.9	
10	11.4	10.9	10.5	10.1	9.9	9.8	9.7	9.6	9.5	9.4	
15	9.7	9.6	9.3	9.2	9.1						
				01	**07**	**02**					
1	19.4	17.9	16.4	15.4	14.7	14.0	13.5	13.1	12.7	12.4	11.9
2	16.9	15.7	14.9	14.2	13.6	13.2	12.8	12.5	12.2	12.0	11.5
3	15.7	14.7	14.0	13.4	12.9	12.7	12.4	12.2	12.0	11.9	11.4
4	15.0	13.9	13.2	12.7	12.4	12.2	12.0	11.9	11.7	11.6	11.1
5	13.9	13.0	12.5	12.1	11.9	11.8	11.6	11.5	11.4	11.3	10.9
7	12.8	12.1	11.6	11.4	11.2	11.1	11.0	11.0	10.9	10.8	
10	11.7	11.4	11.0	10.8	10.7	10.6	10.6	10.6	10.5	10.5	
15	10.1	10.1	10.1	10.2	10.2						
				08	**08**	**02**					
1	23.9	21.1	19.1	17.7	16.7	15.9	15.2	14.6	14.1	13.6	12.9
2	18.9	17.1	16.2	15.3	14.7	14.2	13.9	13.5	13.2	12.9	12.4
3	16.3	15.3	14.6	14.0	13.6	13.3	13.1	12.9	12.7	12.5	11.9
4	15.1	14.2	13.5	13.0	12.7	12.6	12.4	12.2	12.1	12.0	11.5
5	13.9	13.3	12.8	12.4	12.2	12.0	11.9	11.8	11.7	11.6	11.1
7	12.7	12.4	12.0	11.7	11.5	11.4	11.3	11.2	11.2	11.1	
10	11.9	11.6	11.3	11.1	10.9	10.8	10.8	10.8	10.7	10.6	
15	10.4	10.6	10.3	10.3	10.3						

TABLE 3.9:   Initial Libor rates (%)

Lib. Mat. (yr)	14.05.02	03.06.02	01.07.02	08.08.02
0	3.691	3.671	3.560	3.359
0.5	4.425	4.383	3.968	3.400
1	4.984	4.877	4.508	3.932
1.5	5.292	5.219	4.719	4.057
2	5.268	5.167	4.884	4.498
2.5	5.426	5.376	5.052	4.713
3	5.565	5.406	5.301	4.915
3.5	5.664	5.543	5.439	5.090
4	5.852	5.666	5.398	5.203
4.5	5.920	5.793	5.465	5.313
5	5.965	5.866	5.686	5.461
5.5	6.029	6.004	5.781	5.572
6	5.981	5.955	5.874	5.611
6.5	5.964	6.021	5.882	5.663
7	6.062	6.016	6.017	5.785
7.5	6.048	6.071	6.023	5.839
8	6.278	6.051	5.990	5.579
8.5	6.238	6.104	5.988	5.567
9	6.021	6.076	5.970	5.774
9.5	6.004	6.116	5.979	5.776
10	6.136	6.122	5.990	5.763
10.5	6.259	6.118	5.994	5.766
11	6.334	6.127	5.992	5.769
11.5	6.349	6.152	5.983	5.776
12	6.307	6.192	5.969	5.785
12.5	6.259	6.230	5.953	5.793
13	6.217	6.260	5.937	5.799
13.5	6.187	6.282	5.924	5.804
14	6.169	6.294	5.912	5.807
14.5	6.167	6.296	5.905	5.809
15	6.180	6.288	5.901	5.811
15.5	6.193	6.275	5.895	5.809
16	6.204	6.260	5.886	5.804
16.5	6.211	6.243	5.875	5.795
17	6.214	6.223	5.860	5.782
17.5	6.213	6.202	5.842	5.765
18	6.208	6.179	5.821	5.745
18.5	6.199	6.154	5.797	5.721
19	6.184	6.128	5.770	5.692
19.5	6.165	6.101	5.739	5.660

TABLE 3.10:   Stripped caplet vols. (%)

Capl. Mat. (yr)	14.05.02	03.06.02	01.07.02	08.08.02
0.5	17.80	18.50	18.70	23.50
1	17.76	18.54	18.93	24.52
1.5	17.61	18.00	18.96	23.11
2	17.28	17.18	18.79	20.55
2.5	16.63	16.41	18.25	18.58
3	15.62	15.95	17.30	18.14
3.5	14.82	15.45	16.47	17.70
4	14.56	14.74	16.01	16.49
4.5	14.49	14.25	15.44	15.64
5	14.42	14.16	14.65	15.57
5.5	14.21	14.06	14.36	15.31
6	13.79	13.78	14.72	14.54
6.5	13.34	13.62	14.80	14.03
7	12.94	13.60	14.31	14.00
7.5	12.69	13.47	13.73	13.90
8	12.68	13.12	13.31	13.50
8.5	12.80	12.77	13.19	13.24
9	12.95	12.46	13.32	13.20
9.5	12.82	12.23	13.12	13.13
10	12.35	12.07	12.43	13.03
10.5	11.88	11.94	11.82	12.93
11	11.55	11.85	11.48	12.81
11.5	11.40	11.78	11.49	12.66
12	11.43	11.74	11.86	12.47
12.5	11.50	11.69	12.67	12.13
13.5	11.58	11.66	13.00	11.98
14	11.59	11.63	13.25	11.84
14.5	11.54	11.59	13.38	11.73
15	11.45	11.53	13.38	11.63
15.5	11.33	11.47	13.28	11.54
16	11.18	11.39	13.07	11.45
16.5	10.99	11.28	12.73	11.37
17	10.76	11.16	12.25	11.29
17.5	10.48	11.00	11.60	11.21
18	10.15	10.81	10.77	11.13
18.5	9.76	10.58	9.74	11.05
19	9.30	10.31	8.49	10.97
19.5	8.77	9.99	7.00	10.89

we take in (2.21) a scalar volatility function g according to (2.25) and a three parametric initial correlation structure $\rho^{(0)}$ given by (2.46). We carry out a LSq calibration procedure for each data set and proceed in a sequential way. First we calibrate to the first row of the swaption volatility matrix, then to the first two rows, and so on, up to the whole matrix.

In order to demonstrate the instability of straightforward least squares calibration, we run Algorithm 3.1 for 10000 Halton points and 10 local searches, where the search is first restricted to $\rho_\infty > 0.5$ and then $\rho_\infty < 0.5$. The results are listed in Table 3.11, 3.12 and speak for themselves. We clearly see that for each data set and each calibration, we can find comparable accurate LSq fits in completely different parameter regions! Needless to say that such a situation, where for instance the calibrated ρ_∞ is flying around, is highly undesirable in practice. In the next section we will introduce a regularisation procedure which resolves such problems.

3.4   Regularisation via a Collateral Market Criterion

3.4.1   Market Swaption Formula (MSF)

As the swap rate is in fact a weighted sum of forward Libors (1.36) and since in practice the variability of the weight coefficients is small compared to the variability of the Libors, one expects that in a rough approximation the variance of a forward swap rate is given by

$$\text{Var}(S_{p,q}(T_p)) \sim \sum_{i,j=p}^{q-1} w_i^{p,q} w_j^{p,q} \text{Cov}(L_i(T_p), L_j(T_p)), \qquad (3.13)$$

where the weights $w_i^{p,q}$, $p \le i < q$, are considered to be constant over time. By next assuming (although inconsistently) a driftless Black Scholes model for $S_{p,q}$ and L_i, $p \le i < q$, with market quoted swaption volatility $\hat{\sigma}_{p,q}^B$, and caplet volatilities $\hat{\sigma}_i^B$, $p \le i < q$, respectively, we can imply the involved variances in (3.13) from these quotes and thus obtain an "implied version" of approximate relationship (3.13),

$$S_{p,q}^2(0)(\hat{\sigma}_{p,q}^B)^2 \sim \sum_{i,j=p}^{q-1} w_i^{p,q} w_j^{p,q} L_i(0) L_j(0)\, \hat{\sigma}_i^B \hat{\sigma}_j^B \text{Cor}_{L_i(T_p), L_j(T_p)}. \quad (3.14)$$

In fact, (3.14) involves an additional approximation: For two correlated Black–Scholes models S_1 and S_2, with volatility σ_1 and σ_2 respectively, driven by two Brownian motions with correlation ρ, we have $\text{Cov}(S_1(T), S_2(T)) \approx S_1(0)S_2(0)\sigma_1\sigma_2\rho T$, provided that T is not too large (see Appendix A.3.1).

TABLE 3.11:   Sequential LSq calibration with restricted correlations

To Mat. (yr) →	1	2	3	4	5	7	10	15
$\rho_\infty > 0.5$			14	05	02			
η_1	0.23	0.07	0.07	0.07	0.07	0.13	0.13	0.00
η_2	0.30	0.20	0.18	0.18	0.19	0.17	0.17	0.00
ρ_∞	0.59	0.59	0.59	0.59	0.60	0.65	0.65	0.62
a	0.19	0.01	0.03	0.01	0.01	0.47	0.50	0.00
b	0.82	0.66	0.74	0.72	0.76	1.11	1.07	0.62
g_∞	0.50	0.45	0.45	0.44	0.43	0.46	0.46	0.41
RMS (%)	0.63	1.65	1.55	1.76	2.23	2.53	2.94	3.91
$\rho_\infty < 0.5$								
η_1	1.51	0.25	0.25	0.62	0.16	0.00	0.53	0.00
η_2	0.00	0.53	0.53	0.26	0.48	0.00	0.08	0.00
ρ_∞	0.22	0.28	0.28	0.28	0.49	0.50	0.34	0.15
a	0.88	0.66	0.49	0.81	0.22	0.18	2.49	1.86
b	2.17	1.73	1.63	1.79	1.06	0.89	1.83	2.39
g_∞	0.88	0.60	0.57	0.62	0.48	0.45	0.76	0.65
RMS (%)	0.53	1.77	1.77	2.02	2.33	2.53	3.35	4.39
$\rho_\infty > 0.5$			03	06	02			
η_1	0.25	0.13	0.13	0.13	0.00	0.07	0.00	0.02
η_2	0.10	0.17	0.17	0.17	0.00	0.18	0.00	0.00
ρ_∞	0.51	0.65	0.65	0.65	0.57	0.60	0.6	0.61
a	0.00	0.07	0.11	0.16	0.01	0.00	0.00	0.01
b	0.56	0.71	0.78	0.84	0.69	0.65	0.57	0.57
g_∞	0.49	0.46	0.46	0.46	0.44	0.43	0.42	0.41
RMS (%)	0.72	1.22	1.23	1.25	1.59	2.39	2.68	4.23
$\rho_\infty < 0.5$								
η_1	1.37	1.29	0.17	0.25	0.25	0.16	0.15	0.00
η_2	0.18	0.18	0.50	0.53	0.53	0.46	0.18	0.00
ρ_∞	0.21	0.21	0.47	0.28	0.28	0.49	0.18	0.22
a	0.72	0.44	0.03	0.48	0.62	0.08	1.55	0.06
b	2.08	2.26	0.81	1.63	1.71	0.81	2.18	0.84
g_∞	0.84	0.72	0.50	0.57	0.57	0.47	0.66	0.52
RMS (%)	0.73	1.16	1.13	1.40	1.91	2.48	3.17	4.51

TABLE 3.12: Sequential LSq calibration with restricted correlations

To Mat. (yr) →	1	2	3	4	5	7	10	15
$\rho_\infty > 0.5$			**01**	**07**	**02**			
η_1	0.24	0.23	0.07	0.23	0.07	0.07	0.08	0.13
η_2	0.31	0.30	0.20	0.30	0.18	0.18	0.14	0.17
ρ_∞	0.53	0.59	0.60	0.59	0.59	0.59	0.51	0.65
a	2.27	0.61	0.47	0.53	1.72	1.80	0.97	0.00
b	1.50	1.20	1.19	1.28	1.69	1.67	1.41	0.63
g_∞	0.95	0.61	0.55	0.57	0.67	0.67	0.60	0.46
RMS (%)	0.80	2.59	2.49	2.35	2.49	3.06	3.20	6.03
$\rho_\infty < 0.5$								
η_1	0.77	0.61	0.18	0.24	0.49	0.31	0.31	0.21
η_2	0.20	0.32	0.05	0.00	0.20	0.11	0.11	0.23
ρ_∞	0.37	0.35	0.37	0.49	0.45	0.49	0.49	0.32
a	1.88	1.34	2.53	2.51	3.95	3.52	3.24	0.00
b	2.01	1.85	2.11	1.86	2.26	2.01	1.82	1.40
g_∞	1.00	0.83	0.85	0.83	1.00	0.92	0.92	0.49
RMS (%)	1.37	2.64	2.60	2.40	2.64	3.18	3.36	6.24
$\rho_\infty > 0.5$			**08**	**08**	**02**			
η_1	0.37	0.22	0.18	0.09	0.18	0.18	0.03	0.01
η_2	0.00	0.23	0.22	0.27	0.22	0.22	0.08	0.00
ρ_∞	0.69	0.64	0.62	0.69	0.62	0.62	0.57	0.95
a	0.43	1.11	0.47	0.80	0.47	0.47	0.05	0.15
b	0.84	1.45	1.28	1.47	1.25	1.07	0.87	0.75
g_∞	0.76	0.83	0.67	0.66	0.62	0.62	0.53	0.48
RMS (%)	0.77	1.92	2.80	3.05	3.59	3.93	4.11	4.11
$\rho_\infty < 0.5$								
η_1	0.17	0.17	0.17	0.17	0.56	0.56	0.18	0.18
η_2	0.26	0.26	0.26	0.26	0.24	0.24	0.05	0.05
ρ_∞	0.45	0.45	0.45	0.45	0.44	0.44	0.37	0.37
a	1.28	0.83	0.82	0.70	3.54	2.35	2.46	2.24
b	1.56	1.58	1.79	1.68	2.82	2.07	2.18	2.14
g_∞	1.00	0.83	0.75	0.71	1.00	0.97	0.89	0.85
RMS (%)	1.04	2.06	2.93	3.24	3.82	4.13	4.32	4.53

In principle, the global correlations $\text{Cor}_{L_i(T_p), L_j(T_p)}$ could be implied from the cap/swaption market via (3.14). However, solving the correlations from (3.14) breaks down in practice because, apart from numerical stability problems, its solution is usually not consistent with a correlation structure (e.g., "correlations" may be larger than one etc.). Nevertheless, via interpreting (3.14) in another way, namely with respect to a certain given Libor market model, we may consider (3.14) as a "rule of thumb" approximation formula for swaptions within this model.

DEFINITION 3.1   **Market Swaption Formula (MSF)** *Let us assume that a certain Libor market model is given. Then, the MSF poses that implied Black volatilities of model swaption prices are approximately given by*

$$(\sigma_{p,q}^{MSF})^2 := \frac{1}{S_{p,q}^2(0)} \sum_{i,j=p}^{q-1} w_i^{p,q} w_j^{p,q} L_i(0) L_j(0) \, \sigma_i^B \sigma_j^B \, \text{Cor}_{L_i(T_p), L_j(T_p)}, \quad (3.15)$$

where σ_i^B, $p \leq i < q$, are the model consistent caplet volatilities and where $\text{Cor}_{L_i(T_p), L_j(T_p)}$, $p \leq i \leq j < q$, are the global model correlations of the Libor process.

Essentially, the ideas behind formula (3.15) originate from Rebonato (1996) and also Schoenmakers & Coffey (1998). We now propose to regularise joint calibration to caps and swaptions by collateral use of the MSF in the calibration procedure. The key idea is basically as follows: *Calibrate the Libor market model such that the prices of caps and swaptions are fitted "as good as possible" while the MSF formula is matched "reasonably".* This idea is implemented as a modification of the standard mean-squares objective function (3.6) by the Market Swaption Formula. The details are given in Section 3.4.2. By incorporation of the MSF it turns out possible to identify less ambiguously (de-)correlations in the market model. In other words, the MSF serves as an instrument which decides more or less the trade-off between the explanation power of the correlation structure and the scalar volatilities and as such is a remedy for the intrinsic instability of straightforward least squares calibration.

3.4.2   Incorporation of the MSF in the Objective Function

In Section 3.2.2 we simulated artificial data sets of swaption prices by a Libor model (2.21) equipped with volatility function (2.25) and correlation structure (2.46) or (2.47) and found that for typical values of parameters a, b, g_∞ and $\eta_1, \eta_2, \rho_\infty$ the least squares minimization procedure (3.6) returned the input parameters fairly good, though with small RMS errors partially due to the small inaccuracy of the involved swaption approximation. In contrast, from least squares calibration experiments with various Euro-market data in

Section 3.3.2 we found out that for many data sets comparable fits may be achieved by either a rather flat g-function combined with a correlation structure with ρ_∞ relatively close to zero, or a highly non-flat g-function combined with correlations $\rho_{ij} \equiv 1$, hence a one-factor model. This phenomenon particularly occurred in situations where the attainable overall RMS fit was not too good (possibly caused by internal misalignments in the cap/swaption market data). In Section 3.2.1 we explained the cause of this stability problem and here we propose a regularised calibration strategy based on the MSF concept introduced in Section 3.4.1 as a way around.

The MSF (3.15) involves global correlations but there are generally no closed form expressions for global correlations in a Libor market model. However, by neglecting the stochastic nature of the log-Libor drifts (which are in magnitude of second order anyway) it is easy to derive the following approximation in terms of the model factor loadings determined by c_i, g, and ρ,

$$\mathrm{Cor}(L_i(T_p), L_j(T_p)) \approx \mathrm{Cor}(\ln L_i(T_p), \ln L_j(T_p))$$

$$\approx \frac{\int_0^{T_p} g(T_i - t)g(T_j - t)\rho^{(0)}_{i-m(t),j-m(t)}dt}{\sqrt{\int_0^{T_p} g^2(T_i - t)dt}\sqrt{\int_0^{T_p} g^2(T_j - t)dt}}$$

$$=: \rho^{\mathrm{global};\, g,\rho^{(0)}}_{ij,p} . \tag{3.16}$$

Hence, by using approximation (3.16), the MSF (3.15) can be re-expressed in terms of g and $\rho^{(0)}$ by

$$(\sigma^{MSF}_{p,q}(g, \rho^{(0)}))^2 = \frac{1}{S^2_{p,q}(0)} \sum_{i,j=p}^{q-1} w^{p,q}_i w^{p,q}_j L_i(0) L_j(0)\, \sigma^B_i \sigma^B_j\, \rho^{\mathrm{global};\, g,\rho^{(0)}}_{ij,p} . \tag{3.17}$$

We now propose a calibration procedure which may be interpreted as follows:

- *Fit the model to swaption prices by minimizing (3.6) as far as possible, but such that the RMS error corresponding to the "rule-of-thumb" Market Swaption Formula (3.15) is "not too serious"*

- *If an "exact" fit is possible, the calibration procedure should in principle be able to find this fit.*

For implementation of this calibration strategy we propose minimization of the following objective function,

$$RMS(g, \rho^{(0)}) \max\left(RMS(g, \rho^{(0)}), RMS^{MSF}(g, \rho^{(0)})\right), \tag{3.18}$$

where $RMS(g, \rho^{(0)})$ is given in (3.6) and

$$RMS^{MSF}(g, \rho^{(0)}) := \sqrt{\frac{2}{(n-1)(n-2)} \sum_{1 \le p \le q-2,\; q \le n} \left(\frac{\widehat{\sigma}^B_{p,q} - \sigma^{MSF}_{p,q}(g, \rho^{(0)})}{\widehat{\sigma}^B_{p,q}}\right)^2} .$$

The idea behind (3.18) is clear: For parameters g, $\rho^{(0)}$ with $RMS^{MSF}(g; \rho^{(0)})$ $\leq RMS(g; \rho^{(0)})$, the objective function is just equal to the mean squares error $MS(g, \rho^{(0)})$ of the (approximative) model swaption prices with respect to the market quotes, and so disregards the precise value of the MSF fitting error. If $RMS(g, \rho^{(0)}) \leq RMS^{MSF}(g, \rho^{(0)})$, then the objective function equals the geometric mean $\sqrt{MS(g, \rho^{(0)}) \, MS^{MSF}(g, \rho^{(0)})}$, with $MS^{MSF}(g; \rho^{(0)})$ being the mean squares error of the MSF fit. Since search algorithms usually work better with differentiable objects we next replace in (3.18) the function $\max(x, y)$ by $\sqrt[4]{x^4 + y^4}$ which is differentiable for $(x, y) \neq (0, 0)$. Then, we square the objective function and finally obtain the following minimization problem,

$$MS(g, \rho^{(0)}) \sqrt{\{MS(g, \rho^{(0)})\}^2 + \{MS^{MSF}(g, \rho^{(0)})\}^2} \longrightarrow \min_{g, \rho^{(0)}} \quad (3.19)$$

(MS=mean squares), where in our applications the function g and correlation structure $\rho^{(0)}$ are parametrized by (2.25) and (2.46), respectively.

Let us give a further heuristic motivation for the objective function in (3.19). As we argued in Section (3.2.1) the main source of instability in a direct least squares calibration procedure is, roughly speaking, the fact that two different influential objects (scalar volatility function and correlation structure) have to determine one angle, which rotates the model swaption volatility surface to the market volatility surface. Let us now imagine that the MSF formula would be the "correct" swaption pricing formula and that we would try to calibrate this formula by least squares to the market volatilities. Since the MSF returns for $q = p + 1$ the caplet volatilities also, it is obvious from the same arguments that we would encounter similar stability problems. The reason that we expect in practice the objective function (3.19) to be a better candidate for finding an isolated global minimum comes from the following heuristic motivation.

Let the functions $F_1, F_2 : R^2 \to R$ with $F_1, F_2 \geq 0$ be such that

$$F_1(x, y) = 0 \quad \Leftrightarrow \quad y = \varphi_1(x),$$
$$F_2(x, y) = 0 \quad \Leftrightarrow \quad y = \varphi_2(x),$$

where $y = \varphi_1(x) \wedge y = \varphi_2(x) \Leftrightarrow (x, y) = (x^*, y^*)$. Next consider for $\epsilon_1, \epsilon_2 \geq 0$ the functions

$$ms := \epsilon_1 + F_1, \quad ms^{msf} := \epsilon_2 + G.$$

Clearly, ms and ms^{msf} have global minimum ϵ_1 and ϵ_2, attained at the curve $y = \varphi_1(x)$ and $y = \varphi_2(x)$, respectively. With x and y symbolising scalar volatility and correlation, we might consider ms and ms^{msf} as stylistic models for the LSq objective functions MS and MS^{MSF}, respectively. Indeed, both functions take their minimum at a whole curve, rather than at a single point. Consider next the objective function (compare with (3.6)),

$$ms\sqrt{ms^2 + (ms^{msf})^2} = (\epsilon_1 + F_1)\sqrt{(\epsilon_1 + F_1)^2 + (\epsilon_2 + F_2)^2}, \quad (3.20)$$

which has a global minimum $\epsilon_1\sqrt{\epsilon_1^2 + \epsilon_2^2}$. If $\epsilon_1 > 0$ (compare with no exact RMS fit possible), then this minimum is attained only at (x^*, y^*). If $\epsilon_1 = 0$ (compare with exact fit possible) the minimum is attained at the whole curve, $y = \varphi_1(x)$, and then the objective function (3.20) is in a sense equivalent with ms (compare with RMS).

As we saw in Section 3.2.2, where we calibrate to Monte Carlo simulated swaption volatilities due to a particular choice of input parameters a, b, g_∞ and $\eta_1, \eta_2, \rho_\infty$, instead of market swaption quotes, if a very close fit is possible with (3.6) then due to the factor $MS(g, \rho^{(0)})$ in front of (3.19), the concerning parameters will be retrieved (as it should be) and the calibration will not be affected by the MSF. However, in practice the usual case is that there is no very accurate fit via (3.6) possible and then the procedure will return the parameters such that $RMS(g, \rho^{(0)})$ is as close as possible to zero while the average error $RMS^{MSF}(g, \rho^{(0)})$ is not too large, in a sense. *In fact, one then might consider the (eventually refined) swaption price formula (1.41)-(1.42) for the calibrated parameter set as a model based correction of the more intuitive Market Swaption Formula (3.15)!*

3.4.3   MSF Calibration Tests, Regularisation of the Volatility Function

In this section we present the results of several calibration tests for correlation structure (2.46) and volatility function (2.25), based on the regularised objective function (3.19) for the Euro market data in Section 3.3.1. In these tests it turned out that basically the same calibration fit can be achieved by taking the two parametric correlation structure (2.47), which is (2.46) with $\eta_2 = 0$. Indeed, side experiments show that correlation behaviour of the structures (2.46) and (2.47) is virtually indistinguishable in the sense that (2.46), for typical values of $\eta_1, \eta_2, \rho_\infty$ in practice, can be fitted rather well by (2.47) for suitable η, ρ_∞. First we have carried out a sequential calibration to caps and an increasing set of swaptions like in Section 3.3.2, for the LMM with five parameters $\eta, \rho_\infty, a, b, g_\infty$. For the minimum search, Algorithm 3.1 is used with 1000 Halton points and 3 local searches. The results are listed in Table 3.13. In Table 3.13 we observe that the two correlation parameters η, ρ_∞ move smoothly from one calibration to the other; however, the behaviour of the volatility parameters a, b, and g_∞, looks rather erratic. Overall, the calibrated functions $g(s)$ in Table 3.13 tend to sweep relatively fast with a high peak near $s = 0$ and then remains virtually constant. In some cases, a and b even tend to explode (we penalized parameter b for exceeding 100.00) and the peak is getting Dirac-delta shaped. From these observations we may conclude that the decorrelation among the forward Libors in the LMM is settled in a stable way by using the MSF regularisation, but the shape of the volatility function is still unstable. In fact, this shape instability of the volatility function g is no wonder and can be explained as follows. Let us think of a known correlation structure $\rho^{(0)}$ and look at the integrals

$$I^g_{i,j,p} := \frac{1}{T_p} \int_0^{T_p} g(T_i - s)g(T_j - s)\rho^{(0)}_{i-m(s),j-m(s)}ds, \qquad p \leq \min(i,j),$$

which provide the ingredients of the objective function (3.19) (or (3.6) for straightforward least squares). It is not difficult to see that for any given smooth function $g^{(0)} > 0$ it is possible to construct a sequence of smooth functions $g^{(n)}$, $n = 1, 2, \ldots$, such that $g^{(n)}(s) = g^{(0)}(s)$ for $s \geq T_1$, while $I^{g^{(n)}} = I^{g^{(0)}}$ and, for instance, $\sup_{[0,T_1]} g^{(n)} \to \infty$, for $n \geq 1$. Hence, the objective function has the same value for the sequence of pairs $(\rho^{(0)}, g^{(n)})$, $n \geq 0$, while the shape of $g^{(n)}$ looks more and more as a Dirac-delta function near $s = 0$ as $n \to \infty$. So, in particular, there is no unique optimal g, but rather a whole family of functions which minimizes objective function (3.19) (or (3.6) for straightforward least squares). As a consequence, the search routine may run into regions where the parametric form g_{a,b,g_∞} is approaching some Dirac-delta shaped form, as observed in Table 3.13. In order to avoid this problem we will include some more "soft information" in the LMM parametrisation.

By watching roughly the shapes of the calibrated functions g in Table 3.13, it seems that the MSF calibration procedure is able to identify that scalar Libor volatilities close to maturity are higher than far away from maturity, but a detailed identification of the volatility function from a given set of market swaption volatilities in terms of the three parameters a, b, and g_∞ is not possible as explained above. Therefore, and in particular in order to prevent the calibration from searching for Dirac-delta shaped g's, we simply fix the time scale of g by the following pragmatic choice of b,

$$b = \frac{1}{\delta}.$$

The choice $b = 1/\delta$ is motivated by the experimentally observed fact that usually time-shift homogeneous calibration of g to caplets via the LSq minimization procedure (2.20) gives rise to either a monotonically decaying g, or a humped shaped g with a peak around the first tenor $s = T_1 = \delta$. Furthermore, after fixing $b = \frac{1}{\delta}$ we remarkably observed that the best calibration fit is obtained with a simple exponentially decaying volatility function, obtained from (2.25) by taking $a = 0$, hence with the one-parameter function $g_{0,1/\delta,g_\infty}$. The calibration results for the LMM with three free parameters $\eta, \rho_\infty, g_\infty$, are given in Table 3.14, where Algorithm 3.1 is used with 1000 Halton points and 3 local searches. Overall we observe in Table 3.14 regular behaviour of the parameters $\eta, \rho_\infty, g_\infty$, while the quality of the RMS fit is comparable with Table 3.13.

TABLE 3.13:   Calibration via MSF, corr. struct. (2.47), a, b, g_∞ free

To Mat. (yr.) $\rightarrow$	1	2	3	4	5	7	10	15
			14	05	02			
η	1.79	1.79	1.96	2.14	2.14	2.15	2.01	1.61
ρ_∞	0.17	0.17	0.12	0.12	0.12	0.12	0.10	0.08
a	3.51	3.61	0.50	0.00	0.00	0.00	0.00	166.39
b	4.47	3.88	44.24	11.10	7.05	5.80	28.96	100.00
g_∞	1.25	1.22	0.37	0.69	0.71	0.69	0.37	0.35
RMS (%)	0.8	1.9	2.4	2.7	3.1	3.3	3.8	4.9
MSF (%)	2.3	3.5	2.6	2.8	3.3	3.6	4.1	5.2
Time (s)	36.4	147.3	258.4	305.3	585.8	1141.4	3134.7	2140.2
Obj. F.	3.2e-8	4.4e-7	5.2e-7	7.7e-7	1.3e-6	1.9e-6	3.3e-6	8.6e-6
			03	06	02			
η	1.79	1.79	1.79	1.95	2.07	2.14	1.63	1.20
ρ_∞	0.17	0.17	0.17	0.12	0.12	0.12	0.08	0.07
a	3.57	3.67	3.65	42.62	52.25	0.00	76.38	0.00
b	4.51	4.16	3.92	100.00	100.00	8.65	100.00	100.00
g_∞	1.25	1.23	1.22	0.36	0.38	0.67	0.31	0.15
RMS (%)	0.7	1.3	1.5	2.0	2.4	3.1	3.6	5.0
MSF (%)	2.4	2.9	3.1	2.2	2.6	3.3	3.8	5.4
Time (s)	27.3	69.9	127.7	594.7	896.1	507.3	2842.8	3668.9
Obj. F.	2.7e-8	1.5e-7	2.1e-7	2.7e-7	5.3e-7	1.4e-6	2.6e-6	9.7e-6
			01	07	02			
η	1.57	1.57	1.52	1.52	1.71	1.71	1.79	1.15
ρ_∞	0.21	0.21	0.22	0.22	0.18	0.18	0.17	0.11
a	1.79	1.62	0.59	0.59	0.00	0.00	3.81	0.00
b	2.74	3.07	2.24	2.09	3.54	2.91	5.14	5.91
g_∞	1.49	1.19	0.95	0.94	0.81	0.79	1.24	0.62
RMS (%)	1.4	2.8	2.8	2.8	3.1	3.6	3.8	6.6
MSF (%)	1.7	3.2	3.5	3.6	3.3	3.9	4.0	6.8
Time (s)	24.1	70.4	375.4	919.7	281.3	539.7	950.3	1269.6
Obj. F.	6.8e-8	9.8e-7	1.1e-6	1.1e-6	1.5e-6	2.6e-6	3.2e-6	2.7e-5
			08	08	02			
η	0.99	1.11	1.23	1.23	1.23	1.42	1.32	1.21
ρ_∞	0.31	0.33	0.29	0.29	0.29	0.24	0.21	0.21
a	0.72	7.23	3.62	3.68	3.53	3.32	185.5	297.77
b	2.37	8.15	8.63	5.32	4.26	3.82	100.00	100.00
g_∞	1.25	1.50	1.13	1.23	1.21	1.49	0.64	0.69
RMS (%)	1.0	2.2	3.2	3.5	4.0	4.4	4.6	4.9
MSF (%)	1.1	2.3	3.2	3.7	4.4	4.4	4.7	5.0
Time (s)	27.9	74.4	123.3	620.4	583.8	920.0	1985.1	2383.9
Obj. F.	1.6e-8	3.3e-7	1.4e-6	2.2e-6	3.9e-6	5.2e-6	6.6e-6	8.1e-6

TABLE 3.14: Calibration via MSF, corr. struct. (2.47), $b = 1/\delta = 2.00$, $a = 0$

To Mat. (yr) $\rightarrow$	1	2	3	4	5	7	10	15
			14	05	02			
η	1.89	1.88	1.88	2.15	2.15	2.03	2.09	1.66
ρ_∞	0.15	0.15	0.15	0.12	0.12	0.13	0.10	0.08
g_∞	0.87	0.8	0.78	0.91	0.87	0.73	0.83	0.81
RMS (%)	1.0	2.0	2.1	2.7	3.1	3.3	3.9	5.0
MSF (%)	1.7	2.9	3.2	2.8	3.3	3.9	4.1	5.2
Time (s)	17.7	44.6	82.3	133.9	273.8	364.5	408	516.8
Obj. F.	3.4e-8	3.8e-7	4.8e-7	8.0e-7	1.4e-6	2.1e-6	3.3e-6	9.0e-6
			03	06	02			
η	1.89	1.89	1.89	1.89	2.14	2.14	1.71	1.38
ρ_∞	0.15	0.15	0.15	0.15	0.12	0.12	0.08	0.08
g_∞	0.86	0.83	0.81	0.79	0.94	0.89	0.86	0.77
RMS (%)	0.9	1.5	1.6	1.7	2.5	3.2	3.7	5.1
MSF (%)	1.7	2.3	2.5	2.7	2.5	3.3	3.8	5.4
Time (s)	25.0	59.8	108.8	189.3	307.4	331.5	490.8	541.9
Obj. F.	2.6e-8	1.2e-7	1.7e-7	2.3e-7	5.6e-7	1.5e-6	2.7e-6	1.0e-5
			01	07	02			
η	1.52	1.68	1.72	1.79	1.79	1.89	1.73	1.22
ρ_∞	0.22	0.17	0.17	0.17	0.17	0.15	0.15	0.12
g_∞	0.94	0.98	0.97	0.97	0.93	0.97	0.88	0.80
RMS (%)	1.7	3.0	3.1	3.0	3.2	3.7	3.9	6.6
MSF (%)	1.8	3.0	3.1	3.1	3.3	3.7	4.0	6.8
Time (s)	17.3	52.5	98.2	136.8	194.0	253.0	318.0	378.4
Obj. F.	1.3e-7	1.2e-6	1.3e-6	1.2e-6	1.6e-6	2.8c-6	3.2e-6	2.8e-5
			08	08	02			
η	1.01	1.21	1.11	1.31	1.45	1.52	1.39	1.25
ρ_∞	0.30	0.30	0.33	0.27	0.23	0.22	0.22	0.21
g_∞	1.05	1.01	0.83	0.93	0.98	0.98	0.89	0.84
RMS (%)	1.1	2.2	3.1	3.5	4.1	4.4	4.6	4.9
MSF (%)	1.2	2.2	3.5	3.6	4.1	4.4	4.7	5.0
Time (s)	16.5	49.4	77.2	113.0	160.5	227.9	310.3	396.2
Obj. F.	2.3e-8	3.5e-7	1.4e-6	2.3e-6	4.0e-6	5.3e-6	6.6e-6	8.1e-6

3.4.4   Calibration of Low Factor Models

As explained in Section 2.3.5 one can transform for any desired number of factors (Brownian motions) d, a full-rank correlation structure into a rank-d structure by Principal Component Analysis. Therefore it is possible to calibrate a d-factor Libor model by the following algorithm.

ALGORITHM 3.2

(I) *Carry out a multi-factor calibration as outlined in Section 3.4.3 and thus identify the correlation matrix $\rho^{(0)}$.*

(II) *Construct by Principal Component Analysis (see Section 2.3.5) an approximation $\tilde{\rho}^{(0)}$ of $\rho^{(0)}$ with rank d.*

(III) *Substitute $\tilde{\rho}^{(0)}$ for $\rho^{(0)}$ in (3.6) and re-calibrate a, g_∞, hence the volatility function $g_{a,1/\delta,g_\infty}$, by (3.6) while keeping $\tilde{\rho}^{(0)}$ fixed. Then, compute the c_i according to (2.19).*

Note that, since the correlation structure $\rho^{(0)}$ is already determined by the full-factor calibration, the re-calibration of g may be done by straightforward least squares. As an illustration, we have carried out a three factor recalibration from the multi-factor calibration in Table 3.14 and the results are given in Table 3.15. In fact, as we observe from Table 3.15, the price for the dimension reduction is a larger violation of the market swaption formula MSF rather than a larger RMS calibration error.

3.4.5   Implied Calibration to Swaptions Only

At some markets it may happen that caps and swaptions are not well in line with each other in the sense that the attainable RMS calibration fit is not acceptable for a trader. Also it is possible that one needs to price an exotic product which only depends on underlying swap rates, for example a Bermudan swaption. In both cases it might be preferable to calibrate a LMM to swaptions only. The simplest way to do this is to take a virtual Libor model with period length equal to the time span between two settlement dates of a swap in the swap market under consideration. For this model the one-period swaptions can be treated as caplets and thus the one period swaption volatilities need to be interpolated and/or extrapolated to a full system of "caplet volatilities" which covers a desired tenor structure. In this context we repeated all experiments of Section 3.4.3 and the results are listed in Table 3.16, 3.17, 3.18. For completeness we also carried out an un-regularised LSq calibration; see Table 3.19. As to be expected, we see from these tables that a much better overall RMS fit is attainable. However, note that also in

TABLE 3.15:   3-Factor LSq-recalibration of g_{a,b,g_∞}, with correlation structures from Table 3.14 PCA-reduced to 3 factors, $b = 1/\delta = 2.00$, a, g_∞ free

To Mat. (yr) $\rightarrow$	1	2	3	4	5	7	10	15
			14	**05**	**02**			
η	1.89	1.88	1.88	2.15	2.15	2.03	2.09	1.66
ρ_∞	0.15	0.15	0.15	0.12	0.12	0.13	0.10	0.08
a	0.40	0.46	0.22	0.01	0.15	0.60	0.68	0.63
g_∞	0.57	0.54	0.50	0.47	0.46	0.50	0.50	0.46
RMS (%)	1.1	2.1	2.3	3.0	3.5	3.9	4.6	5.6
MSF (%)	11.5	12.0	11.9	11.3	11.5	12.1	12.1	13.2
			03	**06**	**02**			
η	1.89	1.89	1.89	1.89	2.14	2.14	1.71	1.38
ρ_∞	0.15	0.15	0.15	0.15	0.12	0.12	0.08	0.08
a	0.17	0.08	0.22	0.57	0.22	0.46	0.74	0.79
g_∞	0.54	0.51	0.50	0.53	0.50	0.49	0.50	0.48
RMS (%)	0.9	1.5	1.8	2.1	3.0	3.9	4.4	5.8
MSF (%)	11.5	11.4	11.3	11.3	10.8	11.2	12.3	13.4
			01	**07**	**02**			
η	1.52	1.68	1.72	1.79	1.79	1.89	1.73	1.22
ρ_∞	0.22	0.17	0.17	0.17	0.17	0.15	0.15	0.12
a	1.74	0.29	0.29	0.24	0.33	0.08	1.01	0.48
g_∞	0.89	0.59	0.59	0.56	0.55	0.51	0.61	0.49
RMS (%)	1.4	3.1	3.2	3.2	3.6	4.3	4.5	7.0
MSF (%)	8.7	9.4	9.3	9.1	9.3	9.5	10.0	12.2
			08	**08**	**02**			
η	1.01	1.21	1.11	1.31	1.45	1.52	1.39	1.25
ρ_∞	0.30	0.30	0.33	0.27	0.23	0.22	0.22	0.21
a	1.09	0.54	0.47	0.27	0.15	0.29	0.75	0.72
g_∞	0.90	0.75	0.66	0.62	0.58	0.59	0.63	0.60
RMS (%)	1.0	2.4	3.3	3.9	4.6	5.0	5.3	5.4
MSF (%)	5.6	6.3	7.7	7.7	8.1	8.4	8.9	9.2

this situation different comparable LSq fits are possible. For instance, a comparison of Table 3.17 with Table 3.19 indicates instability of straightforward least squares calibration again, and so also here the regularisation techniques presented in this chapter provide a sensible remedy.

3.5　Calibration of a Time-Shift Homogeneous LMM

From an economical point of view it may be appealing to consider a time-shift homogeneous Libor market model. In this respect we speak of a time-shift homogeneous model when the volatility of forward Libor L_i is basically a function of time-to-maturity $T_i - t$ only, rather than a general deterministic function of both T_i and t.

3.5.1　Volatility Structure of Hull-White

The piecewise constant volatility model of Hull-White (2.14) is time-shift homogeneous in the sense that the corresponding Libor volatilities are functions of $T_i - m(t)$, hence the number of reset dates between t and T_i. In this model the volatility scalars Λ_i are in principle completely determined by a system of caplet volatilities via (2.15), (2.16). For existence of a solution Λ, however, the caplet volatilities have to satisfy the decay condition (2.17), which is for an equidistant tenor structure equivalent with (2.18). In Figure 3.3 is plotted the so called *all-in caplet variance* $\left(\sigma_i^B\right)^2 T_i$ as a function of caplet maturity T_i for the implied caplet volatilities of the four data sets under consideration. We see that (2.18) is violated for all these data sets, where this violation is particularly strong for the data sets 14.05.02 and 01.07.02, for longer maturities. As a consequence, a direct identification of the volatility vector Λ by (2.18), hence the Hull-White volatility structure (2.14), is not possible for these data.

REMARK 3.3　As a matter of fact, practitioners tend to feel uneasy about caplet volatilities σ_i^B decaying so quickly that $T_i \to \left(\sigma_i^B\right)^2 T_i$ is decreasing over certain intervals, namely, for the following reason. The all-in caplet variance $\left(\sigma_i^B\right)^2 T_i$ can be regarded as a (perhaps rough) estimation of the variance of $\ln L_i(T_i)$. On the other hand, any one factor *Markovian* model for the spot log-Libor,

$$\Delta_i \ln L := \ln L_{i+1}(T_{i+1}) - \ln L_i(T_i) = \cdots \Delta t + \lambda_i \Delta W,$$

with $\Delta_j T := T_{j+1} - T_j$, would give rise to $Var(\ln L_i(T_i)) \approx \sum_{j=1}^{i} \lambda_j^2 \Delta_j T$, which is nondecreasing in i. This indicates that, in practice, behaviour of

TABLE 3.16: Calibration via MSF to swaptions, corr. struct. (2.47), a, b, g_∞ free

To Mat. (yr) →	1	2	3	4	5	7	10
		14	05	02			
η	0.98	1.00	1.02	1.02	1.02	0.98	0.92
ρ_∞	0.36	0.36	0.36	0.36	0.36	0.34	0.33
a	0.26	0.61	0.26	3.16	0.26	1.48	1.65
b	18.45	37.72	25.84	11.28	51.18	10.21	10.27
g_∞	1.28	1.27	1.06	1.23	0.83	1.07	1.01
RMS (%)	0.6	0.6	0.8	0.8	0.8	1.0	1.1
MSF (%)	0.6	0.6	0.8	0.8	0.8	1.0	1.1
Time (s)	3.8	8.4	24.1	33.3	28.4	85.5	110.7
Obj. F.	1.9e-9	2.4e-9	6.3e-9	6.3e-9	5.4e-9	1.4e-8	2.3e-8
		03	06	02			
η	0.93	0.92	0.93	1.05	1.06	0.95	0.91
ρ_∞	0.31	0.31	0.31	0.35	0.35	0.31	0.31
a	0.48	0.51	0.63	0.91	0.49	0.59	0.77
b	1.71	2.48	4.21	8.65	7.37	4.26	4.03
g_∞	1.24	1.21	1.21	1.12	1.08	1.20	1.19
RMS (%)	0.7	0.8	0.9	0.8	0.8	1.1	1.1
MSF (%)	0.9	0.9	0.9	0.8	0.8	1.1	1.1
Time (s)	8.0	14.3	21.1	50.5	68.1	53.9	62.7
Obj. F.	4.2e-9	6.4e-9	1.0e-8	6.6e-9	5.7e-9	1.8e-8	2.1e-8
		01	07	02			
η	0.73	0.70	0.75	0.78	0.78	0.78	0.78
ρ_∞	0.43	0.43	0.43	0.43	0.46	0.46	0.46
a	1.01	0.97	0.78	0.79	0.00	0.81	0.8
b	1.77	2.15	2.39	3.22	9.21	6.76	6.96
g_∞	1.49	1.39	1.37	1.38	1.08	1.13	1.13
RMS (%)	1.0	0.9	0.9	1.0	1.1	1.2	1.2
MSF (%)	1.4	1	1.1	1.1	1.1	1.2	1.2
Time (s)	9.2	16.6	34.7	45.8	133.6	169.4	188
Obj. F.	2.5e-8	9.1e-9	1.2e-8	1.6e-8	1.8e-8	3.2e-8	2.9e-8
		08	08	02			
η	0.25	0.25	0.29	0.29	0.44	0.23	0.23
ρ_∞	0.62	0.62	0.65	0.64	0.63	0.60	0.60
a	0.56	0.56	1.81	1.80	0.34	1.16	1.00
b	2.44	2.03	3.93	3.13	1.49	2.27	2.18
g_∞	1.23	1.20	1.44	1.50	1.20	1.37	1.31
RMS (%)	1.0	0.9	0.8	1.0	1.1	1.2	1.2
MSF (%)	1.1	0.9	0.9	1.0	1.2	1.2	1.2
Time (s)	7.0	12.8	113.4	101.7	152.6	174.5	191.2
Obj. F.	1.8e-8	8.0e-9	7.2e-9	1.4e-8	2.1e-8	3.0e-8	2.6e-8

TABLE 3.17:   Calibration via MSF to swaptions, corr. struct. (2.47), $b = 1/\delta = 1.00$, $a = 0$

To Mat. (yr) $\rightarrow$	1	2	3	4	5	7	10
		14	05	02			
η	0.94	0.85	0.97	0.98	0.98	0.98	0.88
ρ_∞	0.39	0.30	0.38	0.38	0.38	0.38	0.30
g_∞	0.95	1.03	0.97	0.96	0.95	0.94	1.03
RMS (%)	0.7	0.9	0.8	0.8	0.8	1.0	1.2
MSF (%)	1.0	1.0	1.0	1.0	0.9	1.2	1.2
Time (s)	2.8	6.1	9.1	13.5	18.3	26.0	28.3
Obj. F.	6.1e-9	9.5e-9	7.6e-9	7.4e-9	6.5e-9	1.8e-8	2.9e-8
		03	06	02			
η	0.90	0.90	0.98	0.98	0.98	0.94	0.91
ρ_∞	0.30	0.30	0.38	0.38	0.38	0.30	0.30
g_∞	1.04	1.02	0.93	0.94	0.93	1.04	1.02
RMS (%)	0.8	0.9	0.9	0.8	0.8	1.1	1.1
MSF (%)	1.0	0.9	1.2	1.1	1.1	1.1	1.1
Time (s)	3.2	6.4	8.2	13.2	17.0	24.2	26.3
Obj. F.	8.1e-9	8.2e-9	1.1e-8	1.0e-8	8.9e-9	2.0e-8	2.3e-8
		01	07	02			
η	0.79	0.76	0.79	0.79	0.79	0.79	0.79
ρ_∞	0.46	0.43	0.46	0.46	0.46	0.46	0.46
g_∞	1.04	1.04	1.03	1.02	1.02	1.00	1.00
RMS (%)	1.2	0.9	0.9	0.9	1.1	1.2	1.2
MSF (%)	1.3	1.1	0.9	1.0	1.1	1.2	1.2
Time (s)	2.4	3.5	7.2	10.6	15.8	21.7	23.8
Obj. F.	3.0e-8	1.2e-8	9.3e-9	1.2e-8	1.8e-8	3.3e-8	2.9e-8
		08	08	02			
η	0.23	0.23	0.38	0.38	0.38	0.38	0.38
ρ_∞	0.61	0.61	0.68	0.65	0.64	0.61	0.61
g_∞	1.02	1.00	1.01	1.01	1.00	1.01	1.02
RMS (%)	1.1	0.9	0.8	1.0	1.1	1.2	1.2
MSF (%)	1.1	0.9	0.8	1.0	1.1	1.2	1.2
Time (s)	2.1	3.6	6.2	11.4	14.2	18.1	21.9
Obj. F.	2.1e-8	9.3e-9	7.0e-9	1.4e-8	2.2e-8	3.0e-8	2.7e-8

TABLE 3.18: 3-Factor LSq-recalibration of g_{a,b,g_∞}, with correlation structures from Table 3.17 PCA-reduced to 3 factors, $b = 1/\delta = 1.00$, a, g_∞ free

To Mat. (yr) →	1	2	3	4	5	7	10
		14	**05**	**02**			
η	0.94	0.85	0.97	0.98	0.98	0.98	0.88
ρ_∞	0.39	0.30	0.38	0.38	0.38	0.38	0.30
a	0.00	0.00	0.00	0.00	0.00	0.00	0.00
g_∞	0.74	0.72	0.71	0.71	0.70	0.69	0.69
RMS (%)	0.5	1.3	1.2	1.3	1.3	1.5	1.8
MSF (%)	5.3	5.5	5.2	5.1	5.0	5.1	5.0
		03	**06**	**02**			
η	0.90	0.90	0.98	0.98	0.98	0.94	0.91
ρ_∞	0.30	0.30	0.38	0.38	0.38	0.30	0.30
a	0.00	0.00	0.00	0.00	0.00	0.00	0.00
g_∞	0.73	0.71	0.71	0.70	0.69	0.70	0.69
RMS (%)	0.9	1.1	1.0	1.1	1.2	1.6	1.7
MSF (%)	5.6	5.6	5.5	5.3	5.2	5.0	5.1
		01	**07**	**02**			
η	0.79	0.76	0.79	0.79	0.79	0.79	0.79
ρ_∞	0.46	0.43	0.46	0.46	0.46	0.46	0.46
a	0.04	0.01	0.01	0.04	0.04	0.12	0.12
g_∞	0.83	0.80	0.80	0.80	0.79	0.81	0.81
RMS (%)	1.0	0.9	1.0	1.2	1.4	1.7	1.6
MSF (%)	3.4	3.5	3.5	3.5	3.5	3.6	3.5
		08	**08**	**02**			
η	0.23	0.23	0.38	0.38	0.38	0.38	0.38
ρ_∞	0.61	0.61	0.68	0.65	0.64	0.61	0.61
a	0.13	0.06	0.11	0.06	0.13	0.06	0.06
g_∞	0.92	0.88	0.93	0.89	0.90	0.87	0.86
RMS (%)	1.0	0.9	0.9	1.1	1.3	1.4	1.4
MSF (%)	2.3	2.4	1.9	2.1	2.3	2.4	2.3

TABLE 3.19:   LSq calibration to swaptions only, corr. struct. (2.47), a, b, g_∞ free

To Mat. (yr) $\rightarrow$	1	2	3	4	5	7	10
		14	05	02			
η	1.02	1.02	0.89	0.27	0.92	0.27	0.27
ρ_∞	0.36	0.36	0.40	0.50	0.40	0.50	0.50
a	1.14	3.57	4	3.08	2.79	2.85	2.88
b	8.66	11.94	2.51	2.10	2.48	2.09	2.07
g_∞	1.41	1.51	2.06	1.23	1.68	1.18	1.18
RMS (%)	0.6	0.6	0.8	0.8	0.8	0.9	0.9
MSF (%)	0.7	0.7	1.4	5.7	1.3	5.3	5.2
Time (s)	1.5	4.3	6.9	9.9	14.2	16.7	14.8
		03	06	02			
η	1.18	0.75	0.27	0.83	0.77	0.27	0.27
ρ_∞	0.31	0.40	0.50	0.40	0.40	0.50	0.50
a	0.60	2.22	2.95	2.32	2.66	3.04	2.97
b	1.80	1.91	1.91	2.21	2.35	2	1.92
g_∞	1.50	1.50	1.25	1.50	1.50	1.23	1.25
RMS (%)	0.5	0.7	0.7	0.8	0.8	0.9	0.9
MSF (%)	2.0	2.5	6.3	1.9	2.1	5.5	5.3
Time (s)	2.0	6.0	10.4	13.1	20.2	27.5	32.3
		01	07	02			
η	0.78	0.87	0.25	0.25	0.25	0.61	0.61
ρ_∞	0.42	0.42	0.78	0.77	0.78	0.54	0.54
a	0.56	0.29	1.45	0.82	0.84	0.04	0.04
b	5.86	3.20	1.38	1.29	1.34	0.94	0.97
g_∞	1.49	1.39	1.02	0.86	0.86	0.88	0.88
RMS (%)	1.1	0.8	0.8	0.8	0.9	1.2	1.2
MSF (%)	1.7	1.6	6.4	6.1	6.0	2.1	2.0
Time (s)	1.2	5.4	10.8	17.3	25.8	26.6	30.7
		08	08	02			
η	0.25	0.37	0.25	0.25	0.25	0.14	0.22
ρ_∞	0.61	0.69	0.67	0.70	0.73	0.62	0.76
a	0.56	0.05	0.56	0.56	0.56	1.38	0.23
b	2.51	0.32	2.05	1.56	1.34	3.23	0.87
g_∞	1.26	0.96	1.13	1.11	1.09	1.15	0.94
RMS (%)	1.0	0.8	0.8	1.0	1.0	1.2	1.1
MSF (%)	1.2	1.0	1	1.5	1.8	1.5	2.3
Time (s)	1.8	9.0	6.4	10.7	25.0	20.9	37.7

market caplet volatilities implies forward log-Libor variances which are usually not consistent with a Markovian model for the spot (log-)Libor rate process, hence a model which can be implemented on a tree. By considering the spot Libor as a proxy for the short rate, this phenomenon points towards a non-Markovian short rate process, which is rather natural in a general HJM framework; see Heath, Jarrow and Morton (1992).   ☐

3.5.2 Quasi Time-Shift Homogeneous Volatility Structure

The quasi time-shift homogeneous (or modified Hull-White) structure (2.21) is basically time-shift homogeneous when $c_i \equiv c$. As an alternative to the regularisation described in Section 3.4 we now consider cap/swaption calibration based on the LSq minimization procedure (2.20) in Section 2.2.5, for the parametric function g given by (2.25). The thus obtained pair (g_{a,b,g_∞}, c), $c > 0$, yields a time-shift homogeneous volatility function $t \to cg_{a,b,g_\infty}(T_i - t)$, for each i, which matches as close as possible the given system of Black caplet volatilities. By next adjusting a system of coefficients c_i by (2.19) we can take the scalar volatility structure $\sigma_i(t) := c_i g_{a,b,g_\infty}(T_i - t)$, which is expected to be "nearly time-shift homogeneous", and matches the market caplets perfectly. As we know, cap(let) prices are independent of the correlation structure and so, after determination of g_{a,b,g_∞} in the above way, we can try to fit a correlation structure, for example (2.46), to a given system of market swaption volatilities via least squares minimization (3.12). For the four data sets under consideration the results are given in Table 3.20 and Table 3.21. Apparently, the calibration results for the dates 01.07.02 and 08.08.02 are rather poor. Even the first swaption row could not be fitted well. From this it seems that pre-specifying the function g via caplets may be too restrictive, in the sense that the remaining explanation power of the correlation structure (2.47) may be not sufficient for a good fit. It might be possible, of course, to improve the time-shift homogeneous calibration by using more flexible correlation structures. However, for such structures nice economical and computational features may be lost. For example, Malherbe (2002) proposes a correlation structure which is a generalization of (2.8) in fact, but does not provide conditions to ensure positivity, nor conditions (such as (2.30)) to entail typical forward rate correlation behaviour (see Section 2.3.2).

TABLE 3.20: Time-shift homogeneous calibration to caps and swaptions

To Mat. (yr) $\rightarrow$			1	15
	14	**05**	**02**	
a	0.27	η_1	0.24	0.24
b	0.54	η_2	0.01	0.01
g_∞	0.40	ρ_∞	0.78	0.78
C_{av}	0.19			
RMS (%)	4.7		2.9	4.9
	03	**06**	**02**	
a	0.69	η_1	0.08	0.11
b	0.91	η_2	0.05	0.02
g_∞	0.49	ρ_∞	0.79	0.68
C_{av}	0.18			
RMS (%)	2.5		2.1	4.6
	01	**07**	**02**	
a	0.20	η_1	0.24	0.24
b	0.31	η_2	0.01	0.01
g_∞	0.31	ρ_∞	0.78	0.78
C_{av}	0.18			
RMS (%)	11.3		4.1	7.5
	08	**08**	**02**	
a	2.78	η_1	0.04	0.04
b	1.62	η_2	0.01	0.01
g_∞	0.49	ρ_∞	0.95	0.95
C_{av}	0.19			
RMS (%)	1.8		16.0	11.1

TABLE 3.21: Time-shift homogeneous calibration to swaptions

To Mat. (yr) →			1	10
	14	**05**	**02**	
a	0.07	η_1	0.01	0.01
b	0.40	η_2	0.01	0.01
g_∞	0.41	ρ_∞	0.98	0.98
C_{av}	0.18			
RMS (%)	1.6		2.3	3.2
	03	**06**	**02**	
a	0.07	η_1	0.01	0.01
b	0.40	η_2	0.01	0.01
g_∞	0.41	ρ_∞	0.98	0.98
C_{av}	0.18			
RMS (%)	2.3		1.8	3.1
	01	**07**	**02**	
a	2.65	η_1	0.01	0.01
b	1.27	η_2	0.01	0.01
g_∞	0.65	ρ_∞	0.98	0.98
C_{av}	0.13			
RMS (%)	2.5		10.9	10.7
	08	**08**	**02**	
a	2.03	η_1	0.01	0.01
b	2.43	η_2	0.01	0.01
g_∞	0.31	ρ_∞	0.98	0.98
C_{av}	0.29			
RMS (%)	1.5		23.8	19.4

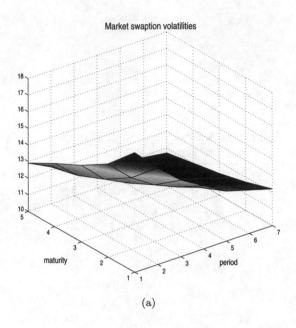

(a)

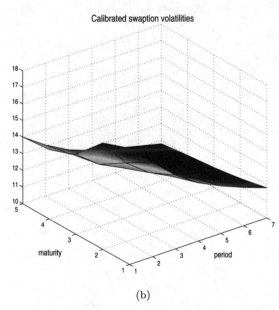

(b)

FIGURE 3.1:   (a) 35 (5x7) Market swaption volatilities, (b) calibrated volatilities

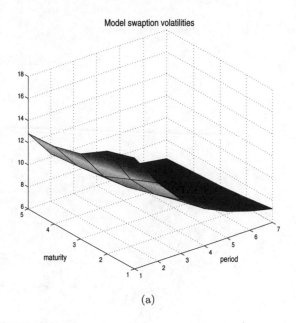

(a)

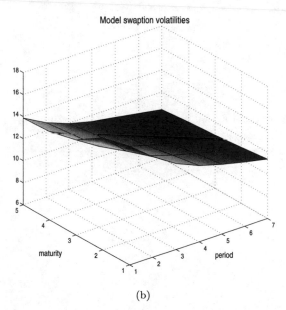

(b)

FIGURE 3.2:   Model swaption vols for two different volatility structures: (a) $a = 0.2$, $b = 1$, $g_\infty = 0.1$; $\eta_1 = \eta_2 = 0$, $\rho_\infty = 0.9$, (b) $a = 0$, $b = 1$, $g_\infty = 0.5$; $\eta_1 = \eta_2 = 0$, $\rho_\infty = 0.1$

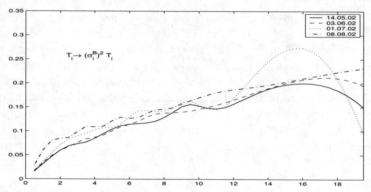

FIGURE 3.3:   All-in caplet variance $\left(\sigma_i^B\right)^2 T_i$ as function of T_i

Chapter 4

Pricing of Exotic European Style Products

4.1 Exotic European Style Products

It will be interesting to see the impact of the different calibration methods studied in Chapter 3 on the prices of different derivative products. In this chapter we consider derivative instruments of European type in a slightly generalized sense. We speak of a European option when the cash-flows at pre-specified exercise dates (standard European product), or at a random date (stopping time) which is explicitly specified in the option contract.[1] Clearly, all European style products in this sense can be evaluated by straightforward Monte Carlo simulation. In the next subsections we introduce some popular exotic derivative products. We will consider all these products with respect to a Libor model in the spot Libor measure given by SDE (2.1). From Section 1.2.1 it is clear how to switch to a Libor SDE in some other numeraire measure.

4.1.1 Libor Trigger Swap

We consider the following version of a trigger swap over a period $[T_p, T_q]$ for a \$1 notional: Up to the first time after T_p, say T_τ, that the spot Libor $L_j(T_j)$ crosses a pre-specified trigger level K_j, the holder receives a spread rate s (or pays when s is negative) over $[T_p, T_\tau)$. Then, at T_τ the holder enters obligatorily into a swap contract with fixed rate κ for the remaining period

---

[1] An American put option, for instance, does not fall into this class since an optimal stopping time depends on the model used for the underlying and thus cannot be specified in the contract.

$[T_\tau, T_q]$. Obviously, the value of the trigger swap at $t = 0$ is given by

$$Trswp(0) := E_*^{\mathcal{F}_0}\Big(\sum_{j=p}^{\tau-1} \frac{s\delta_j}{B_*(T_{j+1})} + \tag{4.1}$$

$$\frac{1}{B_*(T_\tau)} \sum_{j=\tau}^{q-1} \Big(1 - B_q(T_\tau) - \kappa \sum_{j=\tau}^{q-1} \delta_j B_{j+1}(T_\tau)\Big)\Big),$$

in the spot Libor measure P^* (note that $B_*(0) = 1$). In expression (4.1), the spot Libor rolling over account as well as the zero bonds are evaluated exclusively at tenor times and therefore they can be expressed in the Libor process via the formulas

$$B_*(0) := 1, \qquad B_*(T_i) = \frac{1}{B_1(0)} \prod_{p=1}^{i-1} (1 + \delta_p L_p(T_p)), \quad 1 \le i \le n, \tag{4.2}$$

$$B_j(T_i) = \prod_{p=i}^{j-1} (1 + \delta_p L_p(T_i))^{-1}, \quad 1 \le i \le j \le n, \tag{4.3}$$

respectively (see (1.20) for the spot Libor process at an arbitrary time t). So the trigger swap value is estimated via Monte Carlo simulation of the processes $T_i \to B_*(T_i)$, $1 \le i \le n$, and $T_i \to B_j(T_i)$, $1 \le i \le j$, $j = 1, \ldots, n$, via

$$\widehat{Trswp}_M(0) := \frac{1}{M} \sum_{m=1}^{M} \Big(\sum_{j=p}^{\tau-1} \frac{s\delta_j}{B_*^{(m)}(T_{j+1})} + \tag{4.4}$$

$$\frac{1}{B_*^{(m)}(T_\tau)} \sum_{j=\tau}^{q-1} \Big(1 - B_q^{(m)}(T_\tau) - \kappa \sum_{j=\tau}^{q-1} \delta_j B_{j+1}^{(m)}(T_\tau)\Big)\Big).$$

In (4.4), the Monte Carlo sample of spot Libor trajectories $T_i \to B_*^{(m)}(T_i)$, $1 \le i \le n$, and zero bond trajectories $T_i \to B_j^{(m)}(T_i)$, $1 \le i \le j$, $j = 1, \ldots, n$, $m = 1, \ldots, M$, is obtained via (4.2) and (4.3), respectively, from a Monte Carlo simulation of the corresponding Libor process (2.1) in the spot Libor measure P^*.

4.1.2   Ratchet Cap

A ratchet cap with start tenor T_p and terminal tenor T_q is a standard cap over the period $[T_p, T_q]$, except that the strike κ_j for the period between T_j and T_{j+1}, which corresponds to the j-th caplet is the previous Libor plus a spread s, so $\kappa_j = L_{j-1}(T_{j-1}) + s$, for $p < j < q$. The first period strike is defined to be $\kappa_p = s + \kappa_0$, where κ_0 is some exogenously defined strike. The value of the ratchet cap at $t = 0$ is thus given by

$$Rtcap(0) := E_*^{\mathcal{F}_0} \sum_{j=p}^{q-1} \frac{(L_j(T_j) - \kappa_j)^+ \delta_j}{B_*(T_{j+1})}, \qquad (4.5)$$

in the spot Libor measure, where the spot numeraire B_* can be expressed in Libors by (4.2). So a Monte Carlo estimator for the ratchet cap is given by

$$\widehat{Rtcap}(0) := \frac{1}{M} \sum_{m=1}^{M} \sum_{j=p}^{q-1} (L_j^{(m)}(T_j) - \kappa_j)^+ \delta_j \frac{1}{B_*^{(m)}(T_{j+1})}, \qquad (4.6)$$

where M is the number of Monte Carlo trajectories of the Libor process simulated from SDE (2.1).

4.1.3   Sticky Cap

A sticky cap with start tenor T_p and final tenor T_q is a standard cap over the period $[T_p, T_q]$, except that the strike κ_j for the period between T_j and T_{j+1} corresponding to the j-th caplet is the previous capped rate plus a spread s, so $\kappa_j = \min(L_{j-1}(T_{j-1}), \kappa_{j-1}) + s$, for $p < j < q$. As for the ratchet cap, the first period strike is defined to be $\kappa_p = s + \kappa_0$, for some exogenously defined strike κ_0. The price of the sticky cap and the corresponding Monte Carlo estimator is then given by (4.5) and (4.6), respectively, again.

4.1.4   Auto-flex Cap

An auto-flex cap over a period $[T_p, T_q]$ with strike κ and pre-specified exercise number k can be regarded as a standard cap with strike κ over this period, except that only the first k In-The-Money caplets are payed off. The auto-flex cap price at $t = 0$ can thus be represented in the spot Libor measure by

$$Aflcap(0) := E_*^{\mathcal{F}_0} \sum_{j=p}^{q-1} 1_{\{\varkappa_j \leq k\}} (L_j(T_j) - \kappa)^+ \delta_j \frac{1}{B_*(T_{j+1})}, \qquad (4.7)$$

where

$$\varkappa_j := \#\{i : L_i(T_i) > \kappa, \quad p \leq i \leq j\}, \quad p \leq j < q,$$

and the B_* values are expressed in Libors via (4.2). The corresponding Monte Carlo estimator is given by

$$\widehat{Aflcap}(0) := \frac{1}{M} \sum_{m=1}^{M} \sum_{j=p}^{q-1} 1_{\{\varkappa_j^{(m)} \leq k\}} (L_j^{(m)}(T_j) - K)^+ \delta_j \frac{1}{B_*^{(m)}(T_{j+1})}, \qquad (4.8)$$

where M is the number of simulated Monte Carlo trajectories of the Libor process given by SDE (2.1).

4.1.5 Callable Reverse Floater

We now consider the callable reverse floater, an instrument which may behave depending on its particular specification like a floor (cap) or like a swaption (Jamshidian (1997)).

Let $\kappa, \kappa' > 0$. A *reverse floater* (RF) over a period $[T_p, T_q]$ is a contract for receiving $L_i(T_i)$ while paying $\max(\kappa - L_i(T_i), \kappa')$ at time T_{i+1}, for $i = p, .., q-1$, with respect to a unit principle, say \$1. A *callable reverse floater* (CRF) is an option to enter into a reverse floater at T_p. Hence, the CRF will be exercised at T_p when the value of the reverse floater at T_p is positive.

For the reverse floater, the T_{i+1}-cash-flows are thus given by

$$
\begin{aligned}
C_{T_{i+1}} &:= \delta_i L_i(T_i) - \delta_i \max(\kappa - L_i(T_i), \kappa') \\
&= \delta_i(L_i(T_i) - \kappa') - \delta_i \max(\kappa - \kappa' - L_i(T_i), 0).
\end{aligned}
\tag{4.9}
$$

So, by (4.9), the reverse floater can be seen as a swap with strike κ', *minus* a floor with strike $\kappa - \kappa'$. As a consequence, the value of the RF at time t, $t \leq T_p$ can be expressed as

$$
RF(t) = B_p(t) - B_q(t) - \kappa' \sum_{i=p}^{q-1} \delta_i B_{i+1}(t)
$$

$$
- \sum_{i=p}^{q-1} B_{i+1}(t) E_{i+1}^{\mathcal{F}_t} \left[\delta_i \max(\kappa - \kappa' - L_i(T_i), 0) \right]
\tag{4.10}
$$

and therefore can be computed analytically in a Libor market model by Black (and Scholes) type expressions (1.31). The price of the callable reverse floater at time $t = 0$ is next given by

$$
CRF(0) := E_*^{\mathcal{F}_0} \left[\frac{\max(RF(T_p), 0)}{B_*(T_p)} \right],
\tag{4.11}
$$

for which the corresponding Monte Carlo estimator is given by

$$
\widehat{CRF}(0) := \frac{1}{M} \sum_{m=1}^{M} \frac{\max(RF^{(m)}(T_p), 0)}{B_*^{(m)}(T_p)},
$$

for M Monte Carlo trajectories of the Libor process given by SDE (2.1). The zero bonds and spot numeraire involved in (4.11) can be expressed in Libors by (4.2) and (4.3), respectively.

Let us assume for simplicity that $\kappa' = 0$, so $C_{T_{i+1}} := \delta_i L_i(T_i) - \delta_i \max(\kappa - L_i(T_i), 0)$, and consider some special cases. When, for example, $\kappa \approx 2L_i(0)$, the probability that Libors exceed κ within the period $[0, T_q]$ can be neglected in practice, provided the time period $[0, T_q]$ is not too long. So, when the option is called the cash-flows are practically given by $C_{T_{i+1}} := 2\delta_i(L_i(T_i) - \kappa/2)$, and thus the option is basically a swaption on a doubled principal with strike rate $\kappa/2$. If, in contrast, $\kappa \approx L_i(0)$ we may neglect the possibility that Libors fall below $\kappa/2$ and, practically speaking, the cash-flows $C_{T_{i+1}} := \delta_i L_i(T_i) - \delta_i \max(\kappa - L_i(T_i), 0)$ will be surely positive for every i. Then the option will be exercised in any case, yielding a cash-flow equal to the difference of the Libor rate on a forward loan with unit principle and the cash-flow of a floor over the period $[T_p, T_q]$ with strike κ. Therefore, in this situation the valuation of the CRF involves the valuation of a floor. We thus observe that the CRF has *both* cap/floor *and* swaption characteristics. Some more technical details around the callable reverse floater can be found in Schoenmakers & Coffey 1998. In Chapter 6 we consider the valuation of the callable reverse floater again in the context of lognormal Libor approximations.

4.2   Factor Dependence of Exotic Products

An important issue is the impact of a certain kind of Libor model calibration on the price of a particular exotic Libor product. More specifically, the question is: what is the effect of a particular type of calibration, for instance, a two, three, or full factor calibration to a given set of liquid instruments, on the price of a given exotic product. For the different calibration procedures presented in Chapter 3, we thus ask about the sensitivity of a Libor product value with respect to different ways of calibration to the same set of caps and swaptions. As a trivial observation, the members of the calibration set itself, for example the swaptions, are not sensitive as they are matched by the differently calibrated models. As pointed out by Andersen & Andreasen (2000a), also Bermudan swaptions, swaptions which are callable at a time to decide by the contract holder, and studied extensively in Chapter 5, are not very sensitive to the number of factors in the Libor model calibrated to the same set of swaptions. Below, however, we argue that the exotic cap structures introduced in Sections 4.1.2-4.1.4 are quite sensitive to a particular type of calibration. We give a detailed examination for the ratchet cap, from which it will be clear that the same effects apply to the other exotic cap structures. By experiments in Section 4.3 it turns out that the zero-adjusted Libor trigger swap is correlation/factor sensitive as well.

Correlation/factor dependence of the ratchet cap

Let us express the value of the ratchet cap at $t = 0$ in the form

$$
\begin{aligned}
Rtcap(0) &= \sum_{j=p}^{q-1} B_{j+1}(0) E_{j+1}^{\mathcal{F}_0} \, \delta_j (L_j(T_j) - \kappa_j)^+ \\
&= \sum_{j=p}^{q-1} B_{j+1}(0) E_{j+1}^{\mathcal{F}_0} \, (\delta_j L_j(T_j) - \delta_j L_{j-1}(T_{j-1}) - \delta_j s)^+ \\
&=: \sum_{j=p}^{q-1} B_{j+1}(0) E_{j+1}^{\mathcal{F}_0} \, (X_j - \delta_j s)^+ ,
\end{aligned}
\tag{4.12}
$$

and concentrate on two consecutive forward Libors $L_{j-1}(T_{j-1})$ and $L_j(T_j)$ for a fixed j. We first write

$$
\begin{aligned}
X_j &= \delta_j L_j(T_j) - \delta_j L_{j-1}(T_{j-1}) = \delta_j L_j(T_j) \left(1 - \frac{L_{j-1}(T_{j-1})}{L_j(T_j)} \right) \\
&\approx -\delta_j L_j(T_j) \ln \frac{L_{j-1}(T_{j-1})}{L_j(T_j)} \\
&= \delta_j L_j(T_j) \left(\ln L_j(T_j) - \ln L_{j-1}(T_{j-1}) \right),
\end{aligned}
\tag{4.13}
$$

using $1 - z = -\ln z + O\left((z-1)^2\right)$ for $z \longrightarrow 1$. Next, we approximate the joint distribution of $Y_1 := \ln L_{j-1}(T_{j-1})$ and $Y_2 := \ln L_j(T_j)$ with respect to P_{j+1} by a two dimensional normal distribution,

$$
(Y_1, Y_2) \stackrel{d}{\approx} \left(\mu_1 + \sigma_1(\varsigma_1 \sqrt{1-\rho^2} + \varsigma_2 \rho), \mu_2 + \sigma_2 \varsigma_2 \right),
\tag{4.14}
$$

where, in this analysis, μ_1 and μ_2 denote the mean, σ_1^2 and σ_2^2 the variances of Y_1 and Y_2, respectively, ρ denotes the correlation between Y_1 and Y_2, and ς_1, ς_2 are independent standard normal random variables. For a LMM we know that the distribution of Y_2 is exactly normal in P_{j+1} with $\mu_2 = 0$, and, that μ_1 is very small since in the drift of L_{j-1} there is only one term involved. For the present examination we therefore neglect μ_1 and then obtain by (4.13) and (4.14),

$$
X_j \stackrel{d}{\approx} \delta_j e^{\sigma_2 \varsigma_2} \left((\sigma_2 - \rho\sigma_1) \varsigma_2 - \sigma_1 \sqrt{1-\rho^2}\, \varsigma_1 \right).
\tag{4.15}
$$

We next compute and investigate the mean and variance of X_j, approximated by the r.h.s. of (4.15). Note that

$$
E_{j+1}^{\mathcal{F}_0} X_j \approx E_{\varsigma_1, \varsigma_2} X_j = (\sigma_2 - \rho\sigma_1) \, \delta_j E_{\varsigma_2} \, \varsigma_2 e^{\sigma_2 \varsigma_2} = (\sigma_2^2 - \rho\sigma_1\sigma_2)\delta_j e^{\sigma_2^2/2}
\tag{4.16}
$$

and that

$$
\begin{aligned}
E_{j+1}^{\mathcal{F}_0} X_j^2 &\approx E_{\varsigma_1, \varsigma_2} X_j^2 = E_{\varsigma_2} \delta_j^2 e^{2\sigma_2 \varsigma_2} E_{\varsigma_1} \left((\sigma_2 - \rho\sigma_1) \varsigma_2 - \sigma_1 \sqrt{1-\rho^2}\, \varsigma_1 \right)^2 \\
&= \delta_j^2 E_{\varsigma_2} e^{2\sigma_2 \varsigma_2} \left((\sigma_2 - \rho\sigma_1)^2 \varsigma_2^2 + \sigma_1^2(1-\rho^2) \right) \\
&= \sigma_1^2(1-\rho^2)\delta_j^2 e^{2\sigma_2^2} + (\sigma_2 - \rho\sigma_1)^2 \delta_j^2(1 + 4\sigma_2^2)e^{2\sigma_2^2},
\end{aligned}
$$

hence

$$\text{Var}(X_j) := \text{Var}_{\sigma_1,\sigma_2,\rho}(X_j)g \tag{4.17}$$
$$= \sigma_1^2(1-\rho^2)\delta_j^2 e^{2\sigma_2^2} + (\sigma_2 - \rho\sigma_1)^2\,\delta_j^2(1+4\sigma_2^2)e^{2\sigma_2^2} - (\sigma_2^2 - \rho\sigma_1\sigma_2)^2\delta_j^2 e^{\sigma_2^2}.$$

We then consider the derivative with respect to ρ,

$$\frac{\partial}{\partial\rho}\text{Var}_{\sigma_1,\sigma_2,\rho}(X_j) \tag{4.18}$$
$$= \delta_j^2 e^{\sigma_2^2}\left(2\sigma_1\sigma_2^3 - 2\sigma_1\sigma_2 e^{\sigma_2^2} - 8\sigma_1\sigma_2^3 e^{\sigma_2^2} + 2\rho\sigma_1^2\sigma_2^2(4e^{\sigma_2^2}-1)\right),$$

which is linear in ρ with positive slope. Thus, (4.18) may be estimated from above by substituting $\rho = 1$, and then after some rearranging we obtain

$$\frac{\partial}{\partial\rho}\text{Var}_{\sigma_1,\sigma_2,\rho}(X_j) \le -2\sigma_1\sigma_2\delta_j^2 e^{\sigma_2^2}\left(e^{\sigma_2^2} + \sigma_2(\sigma_2 - \sigma_1)(4e^{\sigma_2^2}-1)\right).$$

Note that

$$e^{\sigma_2^2} + \sigma_2(\sigma_2 - \sigma_1)(4e^{\sigma_2^2}-1) \ge e^{\sigma_2^2} - \max(0,\sigma_1-\sigma_2)\,\sigma_2(4e^{\sigma_2^2}-1)$$
$$\ge e^{\sigma_2^2}\left(1 - 4\sigma_2\max(0,\sigma_1-\sigma_2)\right) > 0$$

when $4\sigma_2\max(0,\sigma_1 - \sigma_2) < 1$. Hence,

$$\frac{\partial}{\partial\rho}\text{Var}_{\sigma_1,\sigma_2,\rho}(X_j) \le -2\sigma_1\sigma_2\delta_j^2 e^{2\sigma_2^2}(1 - 4\sigma_2|\sigma_2 - \sigma_1|) < 0$$

for $4\sigma_2\max(0,\sigma_1 - \sigma_2) < 1$, $\rho \le 1$. As a result, when $4\sigma_2\max(0,\sigma_1 - \sigma_2) < 1$ is satisfied, then $\text{Var}_{\sigma_1,\sigma_2,\rho}(X_j)$ decreases in ρ when $\rho \uparrow 1$.

In every Libor market model which is perfectly matched to caplets, we have $\sigma_2 = (\gamma_i^B)^2 T_i$, and $\sigma_1 \approx (\gamma_{i-1}^B)^2 T_{i-1}$ since $\mu_1 \approx 0$ (μ_1 deterministic would be enough for equality), where γ_{i-1}^B and γ_i^B are the respective caplet volatilities. Hence, σ_1 and σ_2 are not affected by a certain kind of calibration method discussed in Chapter 3. But, ρ is of course dependent on the correlation structure of the calibrated model.

In practice the condition $4\sigma_2\max(0,\sigma_1 - \sigma_2) < 1$ comes down to

$$4\gamma_i^B\sqrt{T_i}\max(0,\gamma_{i-1}^B\sqrt{T_{i-1}} - \gamma_i^B\sqrt{T_i}) < 1. \tag{4.19}$$

Clearly, (4.19) is particularly fulfilled for those T_i where $(\gamma_i^B)^2 T_i \ge (\gamma_{i-1}^B)^2 T_{i-1}$, hence the T_i where the all-in caplet variance is increasing; see the discussion in Section 3.5.1. As we can see in Figure 3.3 this is in practice the case for small to mediate term T_i. Otherwise, when $\gamma_{i-1}^B\sqrt{T_{i-1}} = \gamma_i^B\sqrt{T_i}(1+\epsilon)$, for some $\epsilon > 0$, condition (4.19) reads $4\epsilon(\gamma_i^B)^2 T_i < 1$. This roughly means that, if increasingness of the all-in caplet variance is violated, then the violation should not be too strong. Without further discussion we note that, typically, the latter situation applies, at least for the data sets considered in this book.

If we finally approximate X_j with a normal random variable $\widetilde{X}$ with matching mean (4.16) and variance (4.17), we obtain for the j-th term in (4.12) the approximation

$$E_{j+1}^{\mathcal{F}_0} (X_j - \delta_j s)^+ \approx E \left(\widetilde{X} - \delta_j s \right)^+ \qquad (4.20)$$
$$= E \left((\sigma_2^2 - \rho\sigma_1\sigma_2)\delta_j e^{\sigma_2^2/2} + \varsigma\sqrt{\mathrm{Var}_{\sigma_1,\sigma_2,\rho}(X_j)} \right)^+ ,$$

where ς is standard normally distributed. Now, for increasing ρ, the mean term of the Gaussian random variable between the brackets in the r.h.s. of (4.20) shifts into the negative direction, and in the meantime the variance decreases. From this it not difficult to show that the r.h.s. of (4.20) decreases with increasing ρ.

From the above heuristic analysis we conclude that, under some restrictions which are normally fulfilled in practice, the ratchet cap value decreases if the correlations between consecutive forward Libors increase. The latter typically happens when the number of factors in the calibration is decreased! Indeed, if one applies PCA analysis to a full rank correlation structure to reduce the number of factors to two or three, one typically observes that the instantaneous correlations at the first off-diagonal increase and tend to one. In the next section the factor sensitivity of the ratchet cap and other exotic caps is confirmed in different case studies.

REMARK 4.1   A typical effect of a two or three factor Principle Component Analysis applied to a full-rank correlation structure is the fact that correlations $\rho_{i,i+1}$, $\rho_{i,i+2}$, become very close to one while correlations $\rho_{i,i+k}$, for k large, are still comparable with the corresponding correlations of the original input matrix. So the effect of factor reduction should be distinguished from increasing ρ_∞ in a full rank correlation structure. □

4.3   Case Studies

For typical examples of the products introduced in Sections 4.1.1–4.1.4 we here computed Monte Carlo prices with respect to differently calibrated Libor models. In particular, we consider for the data sets of 14.05.02, 03.06.02, 01.07.02, 08.08.02, the respective calibrations given in Tables 3.11–3.15, and price a set of exotic Libor products with respect to each calibration. For each calibrated data set we are going to price the following products by Monte Carlo simulation in the spot Libor measure.

(i) *ATM Libor Cap:* Cap with maturity 1 yr., cap period 1-20 yr., strike equal to the (standard) forward swap rate over this period

(ii) *Auto Flex Cap:* As (i), but only the first 10 ITM caplets are paid off

(iii) *Sticky Cap*

- Case 1: Maturity 1 yr., sticky cap period 1-20 yr., exogenous start-up strike equal to the (standard) forward swap rate over this period, spread 20 bp.

- Case 2: Maturity 1 yr., sticky cap period 1-20 yr., exogenous start-up strike equal to the present Libor rate, spread 30 bp.

(iv) *Ratchet Cap*

- Case 1: Maturity 1 yr., sticky cap period 1-20 yr., exogenous start-up strike equal the present Libor rate, spread 30 bp.

- Case 2: As (iv)-Case 1, but with zero spread

(v) *Libor Trigger Swap* (knock-in)

- Case 1: Maturity 1 yr., period 1-20 yr., trigger level 6.5%, knock-in swap strike 6.5%, spread -30 bp. (where negative means to be paid)

- Case 2: As (v)-Case 1, with spread adjusted such that the contract value is zero

The results are presented in Tables 4.2–4.9. In these tables each particular calibration case is denoted by one of the codes A1/A10–G1/G10, referring to a corresponding calibration case according to Table 4.1.

Discussion of Results

We now discuss the pricing results given in Tables 4.2–4.9, of the above mentioned example products with respect to different calibration methods, calibration instruments, and calibration dates. In this context we address the following issues.

(a) Correlation/factor dependence of different products on a calibration type, with respect to a fixed set of calibration instruments at a fixed calibration date

(b) Dependence of different product prices on the set of calibration instruments, for a fixed calibration method at a fixed calibration date

(c) Dependence of different product prices on the calibration date, for a fixed calibration method and a fixed set of calibration instruments

Add. (a) For each calibration date and set of calibration instruments, 1 yr. or 1–10 yr. maturity swaptions, we conclude the following.

- The prices of the standard cap are consistent with each other, within the Monte Carlo error.

- The Auto-flex-cap coincides overall within the M.C. error on the full-factor calibration cases A1 and C1. For the cases A10 and C10 there is a coincidence within less than 3%, which is of the order of the typical RMS calibration fit for these calibrations. The 3-factor AFC prices due to B1 and B10 are about 5% below A1 and A10 prices, respectively. However, the prices due to calibrations with correlations constraint to $\rho_\infty > 0.5$, hence E1/10, G1/10, are 11-15% below the A1 and A10 prices. We thus conclude that the AFC is a (weakly) correlation sensitive product.

- For the sticky cap, the full-factor prices A1 and C1 differ slightly, 1-2%. The full-factor, LSq calibration with $\rho_\infty < 0.5$, D1,F1, give typically 5-7% below A1/C1. The E1,F1 calibrations ($\rho_\infty > 0.5$) give prices up to 17% below A1,C1. The 3-factor prices B1 fall about 17-20% below A1-C1 prices. For the corresponding 1–10 yr. calibrations, the discrepancies A10–C10 are a bit higher; the other effects are similar. Concluding, we may say that the sticky cap is strongly factor and correlation sensitive.

- The results for the ratchet cap confirm the correlation/factor analysis in Section 4.2. For the chosen examples the 3-factor calibration reduces the A/C full-factor prices even up to 25%. Further, similar conclusions apply as for the sticky cap. In particular, the ratchet cap is very correlation and factor sensitive.

- Also the trigger swap is sensitive for the number of factors and correlations. The relative sensitivities become extremely high of course when the TS is zero-adjusted.

Add. (b) Let us compare for fixed calendar date prices, due to calibrations A1–G1, hence 1 yr. maturity swaptions as calibration instruments, with prices due to A10–G10, 1–10 yr. maturity swaptions as calibration instruments. The caps almost coincide within the Monte Carlo error. The little difference must be caused by different systematic errors due to the SDE simulation. For the exotic caps we observe overall price differences in the order of 3-4% with an exception for the B columns (3-factor calibrations), where the differences are higher. An auxiliary test (not in the table) for trigger swap prices which are not zero-adjusted yielded similar results. Since the RMS calibration fit of A10–G10 is about 3-4% also, we may regard these prices as consistent with the corresponding A1–G1 values, except for the B column. However, since the A1–G1 calibrations are more accurate and the 1 yr. maturity calibration instruments are usually more liquid, the latter calibration procedures may be preferred.

Add. (c) By choosing the strikes of the exotic caps in dependence of the underlying yield curve for the different data sets, these products are more or

TABLE 4.1: Codes for calibration to different instruments at different calibration dates

Calibration Case	
A1/A10:	As in Table 3.14, 1 yr./10 yr. Column
B1/B10:	As in Table 3.15, 1 yr./10 yr. Column
C1/C10:	As in Table 3.13, 1 yr./10 yr. Column
D1/D10:	As in Table 3.11, 1 yr./10 yr. Column, $\rho < 0.5$
E1/E10:	As in Table 3.11, 1 yr./10 yr. Column, $\rho > 0.5$
F1/F10:	As in Table 3.12, 1 yr./10 yr. Column, $\rho < 0.5$
G1/G10:	As in Table 3.12, 1 yr./10 yr. Column, $\rho > 0.5$

TABLE 4.2: Exotic product prices for **X1**-calibrations, 14.05.02, base points (st. dev.)

	14	05	02		
Calibration $\longrightarrow$	A1	B1	C1	D1	E1
Cap	921 (8)	925 (9)	911 (8)	914 (8)	920 (9)
Auto Flex Cap	444 (3)	425 (3)	440 (3)	436 (3)	399 (3)
Sticky Cap 1	1074 (6)	888 (6)	1078 (6)	1050 (6)	939 (6)
Sticky Cap 2	985 (5)	815 (5)	995 (5)	956 (5)	865 (5)
Rachet Cap 1	377 (3)	282 (2)	373 (3)	359 (3)	323 (3)
Rachet Cap 2	531 (3)	434 (2)	537 (3)	517 (3)	473 (3)
Trigger Swap	3 (7)	91 (7)	3 (7)	16 (7)	131 (5)
Trsw. Spread	-30	-30	-30	-30	-30
Zero-Adj. Spread	-30	-50	-30	-35	-60

less standardised. We therefore observe a smooth behaviour of (exotic) cap prices, when going from one datum to the next.

TABLE 4.3:   Exotic product prices for **X10**-calibrations, 14.05.02, base points (st. dev.) (one standard deviation)

| Calibration ⟶ | 14 | 05 | 02 | | |
	A10	B10	C10	D10	E10
Cap	908 (8)	905 (8)	901 (8)	920 (9)	908 (9)
Auto Flex Cap	460 (3)	436 (3)	446 (3)	421 (3)	399 (3)
Sticky Cap 1	1125 (6)	958 (6)	1097 (5)	1023 (6)	946 (6)
Sticky Cap 2	1034 (5)	866 (5)	998 (5)	938 (5)	893 (6)
Rachet Cap 1	405 (3)	317 (3)	387 (3)	353 (3)	336 (3)
Rachet Cap 2	551(3)	463 (2)	541 (3)	513 (3)	487 (3)
Trigger Swap	-23 (7)	54 (7)	-30 (7)	56 (6)	107 (5)
Trsw. Spread	-30	-30	-30	-30	-30
Zero-Adj. Spread	-22	-45	-25	-44	-55

TABLE 4.4:   Exotic product prices for **X1**-calibrations, 03.06.02, base points (st. dev.)

| Calibration ⟶ | 03 | 06 | 02 | | |
	A1	B1	C1	D1	E1
Cap	928 (8)	916 (9)	931 (8)	941 (8)	925 (9)
Auto Flex Cap	447 (3)	426 (3)	441 (3)	431 (3)	407 (3)
Sticky Cap 1	1138 (6)	901 (6)	1106 (6)	1058 (6)	971 (6)
Sticky Cap 2	1001 (5)	806 (5)	983 (5)	956 (5)	874 (5)
Rachet Cap 1	382 (3)	287 (2)	378 (3)	366 (3)	332 (3)
Rachet Cap 2	538 (3)	442 (2)	535 (3)	520 (3)	485 (3)
Trigger Swap	1 (8)	98 (7)	2 (7)	26 (7)	101 (6)
Trsw. Spread	-30	-30	-30	-30	-30
Zero Adj. Spread	-30	-50	-30	-36	-55

TABLE 4.5: Exotic product prices for **X10**-calibrations, 03.06.02, base points (st. dev.)

	03	06	02		
Calibration ⟶	A10	B10	C10	D10	E10
Cap	924 (8)	924 (8)	916 (8)	928 (8)	930 (9)
Auto Flex Cap	461 (3)	437 (3)	456 (3)	428 (3)	397 (3)
Sticky Cap 1	1149 (6)	960 (6)	1114 (5)	1045 (6)	952 (6)
Sticky Cap 2	1026 (5)	860 (5)	1000 (5)	960 (5)	877 (5)
Rachet Cap 1	393 (3)	319 (3)	385 (3)	369 (3)	336 (3)
Rachet Cap 2	555 (3)	473 (3)	539 (3)	527 (3)	488 (3)
Trigger Swap	30 (7)	65 (7)	-16 (7)	44 (6)	115 (6)
Trsw. Spread	-30	-30	-30	-30	-30
Zero-Adj. Spread	-25	-40	-25	-42	-55

TABLE 4.6: Exotic product prices for **X1**-calibrations, 01.07.02, base points (st. dev.)

	01	07	02		
Calibration ⟶	A1	B1	C1	F1	G1
Cap	958 (9)	963 (10)	969 (9)	958 (9)	981 (10)
Auto Flex Cap	459 (4)	442 (4)	465 (4)	442 (4)	422 (4)
Sticky Cap 1	1089 (6)	929 (7)	1113 (6)	1022 (6)	969 (6)
Sticky Cap 2	936 (6)	786 (6)	956 (6)	901 (6)	857 (6)
Rachet Cap 1	353 (3)	249 (2)	353 (3)	326 (3)	298 (3)
Rachet Cap 2	510 (3)	253 (2)	513 (3)	486 (3)	462 (3)
Trigger Swap	-48 (7)	49 (7)	-45 (8)	-2 (7)	49 (6)
Trsw. Spread	-30	-30	-30	-30	-30
Zero-Adj. Spread	-20	-40	-20	-30	-40

TABLE 4.7:   Exotic product prices for **X10**-calibrations, 01.07.02, base points (st. dev.)

| | 01 | 07 | 02 | |
Calibration $\longrightarrow$	A10	B10	C10	F10	G10
Cap	959 (9)	961 (9)	966 (9)	957 (10)	977 (10)
Auto Flex Cap	471 (4)	451 (4)	475 (4)	423 (3)	420 (4)
Sticky Cap 1	1140 (6)	970 (6)	1155 (6)	1040 (6)	1006 (6)
Sticky Cap 2	993 (5)	843 (6)	990 (5)	898 (6)	865 (6)
Rachet Cap 1	387 (3)	291 (3)	389 (3)	329 (3)	325 (3)
Rachet Cap 2	538 (3)	445 (3)	542 (3)	483 (3)	480 (3)
Trigger Swap	-103 (8)	3 (7)	-96 (8)	28 (6)	44 (6)
Trsw. Spread	-30	-30	-30	-30	-30
Zero-Adj. Spread	-10	-30	-10	-35	-37

TABLE 4.8:   Exotic product prices for **X1**-calibrations, 08.08.02, base points (st. dev.)

| | 08 | 08 | 02 | |
Calibration $\longrightarrow$	A1	B1	C1	F1	G1
Cap	995 (10)	977 (10)	981 (10)	991 (10)	997 (10)
Auto Flex Cap	447 (4)	426 (4)	443 (4)	430 (4)	413 (4)
Sticky Cap 1	1053 (7)	936 (7)	1067 (6)	989 (7)	963 (7)
Sticky Cap 2	856 (6)	716 (6)	847 (6)	783 (6)	757 (6)
Rachet Cap 1	307 (3)	227 (2)	306 (3)	277 (3)	266 (3)
Rachet Cap 2	464 (3)	384 (3)	467 (3)	437 (3)	424 (3)
Trigger Swap	-45 (7)	37 (7)	-47 (7)	20 (6)	53 (6)
Trsw. Spread	-30	-30	-30	-30	-30
Zero-Adj. Spread	-25	-35	-23	-32	-38

TABLE 4.9: Exotic product prices for **X10**-calibrations, 08.08.02, base points (st. dev.)

Calibration $\longrightarrow$	08 A10	08 B10	08 C10	02 F10	G10
Cap	997 (9)	986 (10)	1000 (9)	989 (10)	1010 (10)
Auto Flex Cap	453 (4)	446 (4)	456 (4)	433 (4)	413 (4)
Sticky Cap 1	1153 (7)	1001 (7)	1120 (7)	1073 (7)	1019 (7)
Sticky Cap 2	923 (6)	797 (6)	901 (6)	872 (6)	815 (6)
Rachet Cap 1	346 (3)	273 (3)	338 (3)	318 (3)	309 (3)
Rachet Cap 2	501 (3)	434 (3)	494 (3)	482 (3)	478 (3)
Trigger Swap	-106 (7)	-25 (7)	-96 (7)	-30 (6)	12 (5)
Trsw. Spread	-30	-30	-30	-30	-30
Zero-Adj. Spread	-10	-27	-12	-26	-32

Chapter 5

Pricing of Bermudan Style Libor Derivatives

5.1 Orientation

This chapter is devoted to the pricing of Bermudan style Libor derivatives by Monte Carlo methods. A Bermudan Libor derivative is an American option on a pay-off function of forward Libors, which may be called at a finite number of exercise dates. As such we have a high dimensional pricing problem due to the typically high dimension of the underlying Libor process. Generally, evaluation of American style derivatives on a high dimensional system of underlyings is considered a perennial problem for the last decades. On the one hand such high dimensional options are difficult, if not impossible, to compute by PDE methods for free boundary value problems. On the other hand Monte Carlo simulation, which is for high dimensional European options an almost canonical alternative to PDE solving, is for American options highly non-trivial since the (optimal) exercise boundary is usually unknown.

In the past literature, many approaches for Monte Carlo simulation of American options are developed. With respect to Bermudan derivatives, which are in fact American options with a finite number of exercise dates, there is, for example, the stochastic mesh method of Broadie & Glasserman (1997, 2000), a cross-sectional regression approach by Longstaff & Schwartz (2001), and a method by Andersen (1999) for computation of a parametric approximation of the exercise boundary. In general, the price of an American option can be represented as a supremum over a set of stopping times. As a remarkable result, Rogers (2001) (and independently Haugh & Kogan (2001) for Bermudan style instruments) showed that this supremum representation can be converted into a "dual" infimum representation, where the infimum is taken over a set of (super-)martingales. The idea behind this duality approach is rooted in earlier work of Davis & Karatzas (1994) in fact. In Joshi & Theis (2002) duality is used for finding Bermudan swaption prices via a minimization procedure. An alternative (multiplicative) dual approach based on change of numeraires is proposed by Jamshidian (2003,2004). Kolodko & Schoenmakers (2003,2004a) propose a trick for reducing the numerical complexity of computing upper bounds by duality and in Kolodko & Schoenmakers (2004) a general convergent procedure for the optimal Bermudan stopping time is developed. Further recent papers on methods for high-dimensional American options include Belomestny & Milstein (2004), Milstein, Reiß & Schoenmakers (2003), Berridge & Schumacher (2004), and for a more detailed and general overview we refer to Glasserman (2003) and the references therein.

In this chapter we will give a concise recap of the Bermudan pricing problem and describe different Monte Carlo methods for Bermudan style derivatives with applications to Bermudan swaptions. Besides the method of Andersen (1999) and the duality approach of Rogers (2001) and Haugh & Kogan (2001), we present two methods recently developed in Kolodko & Schoenmakers (2003,2004,2004a). The method of Kolodko & Schoenmakers (2003,2004a), exposed in Section 5.8, deals with the problem of efficient numerical evaluation of dual Bermudan upper bounds. A new generic iterative procedure for constructing the optimal Bermudan stopping time, hence the Snell envelope, is developed in Kolodko & Schoenmakers (2004) and here presented in Sections 5.4-5.7. This procedure basically improves a canonical exercise policy for Bermudan options which is already not far from optimal usually; *exercise as soon as the cash-flow dominates all the Europeans ahead.* As such the method has a flavour of "policy iteration" but is rather different from standard policy iteration as in Howard (1960) and Puterman (1994). In particular, the here proposed iterative method does not rely on explicit knowledge of respective transition kernels, is essentially dimension independent, and gives very good results with only a few iterations. These are important features since in various applications, for example, in the case of an underlying (high-dimensional) Libor process transition kernels are not available. Finally, in Section 5.9.1, we deal with multiple callable Bermudans and sketch how the iterative procedure for standard Bermudans can be carried over to such products. For more details we refer to Bender & Schoenmakers (2004). The study of multiple exercise options has recently gained more attraction; see also Carmona & Dayanik (2004), Carmona & Touzi(2004), and Meinshausen & Hambly (2004).

5.2   The Bermudan Pricing Problem

In this section we introduce Bermudan style Libor derivatives with respect to an underlying Libor process in the setting of Theorem 1.3. Let us consider a subset of tenor dates $\mathbb{T} := \{\mathcal{T}_0, \mathcal{T}_1, \ldots, \mathcal{T}_k\} \subset \{0, T_1, \ldots, T_n\}$ with $0 = \mathcal{T}_0 < \mathcal{T}_1 < \mathcal{T}_2 < \cdots < \mathcal{T}_k \leq T_n$. An option issued at time $t = 0$, to exercise once a cash-flow $C_{\mathcal{T}_\tau} := B_n(\mathcal{T}_\tau) c(\mathcal{T}_\tau, L(\mathcal{T}_\tau))$, for some given function $c(\cdot, \cdot)$, at a date $\mathcal{T}_\tau \in \mathbb{T}$ to be decided by the option holder, is called a Bermudan style Libor derivative. We consider the Libor dynamics with respect to a certain pricing measure P connected with some discounting numeraire B, where B is such that $B_n(t)/B(t)$ is measurable with respect to (i.e., is determined by) $L(s)$, $0 \le s \le t$. For example, we can take the dynamics (2.1) in the spot Libor measure where B is the spot Libor rolling-over account (1.20), or SDE (1.16) in the terminal measure where B is taken to be the terminal bond B_n. The value of the Bermudan derivative at time t, $t \ge 0$ (when the option is not exercised before t), is then given by

$$V(t) = B(t) \sup_{\tau \in \{\kappa(t),...,k\}} E_P^{\mathcal{F}_t} \frac{C_{\mathcal{T}_\tau}}{B(\mathcal{T}_\tau)} =: B(t) \sup_{\tau \in \{\kappa(t),...,k\}} E^{\mathcal{F}_t} Z^{(\tau)} \qquad (5.1)$$

with $\kappa(t) := \min\{m : \mathcal{T}_m \ge t\}$, and $Z^{(\tau)}$ denoting the discounted cash-flow at $\mathcal{T}_\tau$. The supremum in (5.1) is taken over all integer valued $\mathbb{F}$-stopping times τ with values in the set $\{\kappa(t),...,k\}$, where $\mathbb{F} := \{\mathcal{F}_t, 0 \le t \le T\}$ denotes the usual filtration generated by the Libor process L (see also Theorem 1.3). For technical reasons it is assumed that $Z^{(i)}$ has finite expectation for each i, $0 \le i \le k$. Note that $V(t)$ can also be seen as the price of a Bermudan option newly issued at time t, with exercise opportunities $\mathcal{T}_{\kappa(t)}, \ldots, \mathcal{T}_k$. The fact that (5.1) can be considered as the fair price for the Bermudan derivative is due to general no-arbitrage principles, e.g., see Duffie (2001).

A family of $\mathbb{F}$-stopping times (indices) $(\tau_t^*)_{0 \le t \le \mathcal{T}_k}$, with $\tau_t^* \in \{\kappa(t), ..., k\}$, is called an *optimal stopping family*, if

$$Y^*(t) := \frac{V(t)}{B(t)} = \sup_{\tau \in \{\kappa(t),...,k\}} , E^{\mathcal{F}_t} Z^{(\tau)} = E^{\mathcal{F}_t} Z^{(\tau_t^*)}, \quad 0 \le t \le \mathcal{T}_k.$$

The thus induced process Y^*, called the *Snell envelope process*, is a super-martingale. Indeed, let (τ_t^*) be an optimal stopping family (which exists by general arguments) and $s < t$. Then it holds

$$E^{\mathcal{F}_s} Y^*(t) = E^{\mathcal{F}_s} E^{\mathcal{F}_t} Z^{(\tau_t^*)} = E^{\mathcal{F}_s} Z^{(\tau_t^*)} \le \sup_{\tau \in \{\kappa(s),...,k\}} E^{\mathcal{F}_s} Z^{(\tau)} = Y^*(s).$$

Henceforth we will consider the process Y^* in particular at the exercise dates $\{\mathcal{T}_0, \mathcal{T}_1, ..., \mathcal{T}_k\}$, and we therefore introduce the discrete process $Y^{*(j)} := Y^*(\mathcal{T}_j)$, $0 \le j \le k$, together with the discrete filtration $\mathcal{F}^{(j)} := \mathcal{F}_{\mathcal{T}_j}$, $0 \le j \le k$. Further we consider a corresponding discrete stopping family (τ_j^*), where $\tau_j^* := \tau_{\mathcal{T}_j}^*$, for $0 \le j \le k$.

Backward Dynamic Programming

It is well known and intuitively obvious that the discrete Snell envelope

$$Y^{*(j)} = E^{\mathcal{F}^{(j)}} Z^{(\tau_j^*)}, \qquad 0 \le j \le k, \qquad (5.2)$$

can be constructed by the following algorithm, called *backward dynamic programming* (see, e.g., Shiryayev (1978), Elliot & Kopp, 1999). At the last exercise date we have $Y^{*(k)} = Z^{(k)}$ and for $0 \leq j < k$,

$$Y^{*(j)} = \max\left(Z^{(j)}, E^{\mathcal{F}^{(j)}} Y^{*(j+1)} \right). \tag{5.3}$$

An optimal stopping family is next represented by

$$\tau_i^* = \inf\left\{ j, i \leq j \leq k : Y^{*(j)} \leq Z^{(j)} \right\}. \tag{5.4}$$

Consider a situation where the discrete process $j \to (L(T_j), B(T_j))$ is Markovian. For instance, consider a Libor market model in the spot Libor numeraire or in the terminal bond measure. Then, obviously, $Y^{*(j)}$ is a function of $L(T_j)$ and $B(T_j)$ for $0 \leq j \leq k$, and thus the conditional expectations in (5.3) may be simulated by Monte Carlo using (regular) conditional probabilities $P^{(L(T_j), B(T_j))}$ (see Appendix A.1.4). So it is possible, in principle, to construct the optimal stopping time via (5.4) while the Snell envelope is constructed backwardly by straightforward Monte Carlo simulation of the backward dynamic program (5.3). But, the complexity of this procedure would be formidable due to its nested structure: N simulations for computing each conditional expectation in (5.3) would lead to $O(N^k)$ samples for the whole set of k exercise dates. For example, $N = 40$ and $k = 10$ would lead to about 40^{10} ($> 10^{16}$) samples!

REMARK 5.1 It is always possible to choose the numeraire B such that $B(T_j)$ is a function of $L(T_j)$ for all j, $0 \leq j \leq m(T_k)$, with $m(\cdot)$ as in (1.20). For instance, one could take for B simply the terminal bond B_n, or any linear combination of $B_{m(T_k)}, \ldots, B_n$. For a Markovian Libor model (e.g., a market model) with respect to such a numeraire the conditional expectations in (5.2) may then be replaced with $E^{L(T_j)}$ rather than $E^{(L(T_j), B(T_j))}$, $0 \leq j \leq k$, as in the case of the spot Libor numeraire. For notational convenience we will therefore assume such kind of numeraires in the description of several Monte Carlo methods in this chapter. It will be obvious, however, how to interpret the presented algorithms, for example, in terms of the spot Libor numeraire which may be more suitable for certain products in practice. ⬜

5.3 Backward Construction of the Exercise Boundary

As a first practically feasible procedure for pricing Bermudan derivatives, we discuss a method for constructing the exercise region of a Bermudan product via backward induction. Obviously, once the exercise region is known, the Bermudan price can be computed by straightforward Monte Carlo simulation.

Let $\mathcal{B}^{(l)}$ be a Bermudan option which is only exercisable at $T_l, ..., T_k$, and τ_l^* be an optimal stopping time for $\mathcal{B}^{(l)}$, for $0 \leq l \leq k$. If $V^{(l)}(0)$ is the value of $\mathcal{B}^{(l)}$ at $t = 0$, we have

$$
\begin{aligned}
\frac{V^{(l)}(0)}{B(0)} &= E^{\mathcal{F}_0} Y^{*(l)} = E^{\mathcal{F}_0} \left(Y^{*(l)} 1_{\tau_l^* > l} + Z^{(l)} 1_{\tau_l^* = l} \right) \\
&= E^{\mathcal{F}_0} \left(1_{\tau_l^* > l} E^{\mathcal{F}^{(l)}} Z^{(\tau_l^*)} \right) + E^{\mathcal{F}_0} \left(Z^{(l)} 1_{\tau_l^* = l} \right) \\
&= E^{\mathcal{F}_0} \left(1_{\tau_l^* > l} E^{\mathcal{F}^{(l)}} Z^{(\tau_{l+1}^*)} \right) + E^{\mathcal{F}_0} \left(Z^{(l)} 1_{\tau_l^* = l} \right) \\
&= E^{\mathcal{F}_0} \left(1_{\tau_l^* > l} Z^{(\tau_{l+1}^*)} \right) + E^{\mathcal{F}_0} \left(Z^{(l)} 1_{\tau_l^* = l} \right).
\end{aligned}
\tag{5.5}
$$

We now assume a Markovian Libor model (e.g., a market model) and the numeraire B to be as in Remark 5.1. For instance, think of B being the terminal bond B_n. Then, due to Markovianity of L, we may replace in (5.5) $\mathcal{F}^{(l)}$ by the sigma algebra generated by $L(T_l)$, and by general arguments there exists for each l, $0 \leq l \leq k$, a smooth function $H_*^{(l)} : R^{n-1} \to R$ which describes the optimal exercise region for the Bermudan $\mathcal{B}^{(l)}$ at T_l by $H_*^{(l)}(L(T_l)) \geq 0$. As a consequence,

$$
\tau_l^* = \inf\{j : j \geq l, \; H_*^{(j)}(L(T_j)) \geq 0\},
\tag{5.6}
$$

so $\{\tau_l^* = l\} = \{H_*^{(l)}(L(T_l)) \geq 0\}$. Since (without loss of generality) cash-flows are assumed to be non-negative, we have $\tau_k^* \equiv k$ and thus we may take, for instance,

$$
H_*^{(k)}(L(T_k)) \equiv 0.
\tag{5.7}
$$

Hence $H_*^{(k)}$ is considered to be trivially known. Equation (5.5) may be expressed in terms of the $H_*^{(l)}$ by

$$
\frac{V^{(l)}(0)}{B(0)} = E^{\mathcal{F}_0} \left(1_{H_*^{(l)}(L(T_l)) < 0} Z^{(\tau_{l+1}^*)} \right) + E^{\mathcal{F}_0} \left(Z^{(l)} 1_{H_*^{(l)}(L(T_l)) \geq 0} \right).
\tag{5.8}
$$

Suppose that for some l, $0 \leq l < k$, the functions $H_*^{(j)}$, $l < j \leq k$, are known (which is by (5.7) the case for $l = k - 1$). Then, the stopping rule τ_{l+1}^* is known by (5.6), and $H_*^{(l)}$ is (formally) a solution of the following optimization problem,

$$
\frac{V^{(l)}(0)}{B(0)} = \max_{H \in \mathcal{H}} E^{\mathcal{F}_0} \left(1_{H(L(T_l)) < 0} Z^{(\tau_{l+1}^*)} \right) + E^{\mathcal{F}_0} \left(Z^{(l)} 1_{H(L(T_l)) \geq 0} \right),
\tag{5.9}
$$

where $\mathcal{H}$ is some suitable class of smooth functions $H : R^{n-1} \to R$. Based on some procedure for the optimization problem (5.9), we may proceed, at least in principle, to construct the whole sequence $H_*^{(l)}$, $0 \leq l < k$, backwardly. This general procedure is utilized by Andersen (1999) in the context of Bermudan swaptions in the Libor market model.

Andersen's Method

In fact, the method of Andersen (1999) can be applied to general Bermudan style derivatives and essentially works as follows. Let us assume that $H_*^{(j)}$ is known for $l < j \leq k$. By (5.7) this is always the case for $l = k - 1$. Consider some rich enough parametric family of functions $H_\alpha^{(l)} : R^{n-1} \to R$, where α runs through some parameter set A. For example, $A \subset R^q$ for some integer q. Then, clearly,

$$\frac{V^{(l)}(0)}{B(0)} \geq \sup_{\alpha \in A} \left(E^{\mathcal{F}_0} \left(1_{H_\alpha^{(l)}(L(T_l)) < 0} Z^{(\tau_{l+1}^*)} \right) + E^{\mathcal{F}_0} \left(Z^{(l)} 1_{H_\alpha^{(l)}(L(T_l)) \geq 0} \right) \right),$$

since each particular α represents a sub-optimal exercise decision. Now consider a Monte Carlo simulation of sample paths $L^{(m)}$, $m = 1, ..., M$, with initial value $L(0)$. Since $H_*^{(j)}$, $l < j \leq k$, is known by assumption, we know the stopping rule τ_{l+1}^* by (5.6), and then we may carry out the optimization

$$\frac{1}{M} \sum_{m=1}^{M} \left(1_{H_\alpha^{(l)}(L^{(m)}(T_l)) < 0} Z_m^{(\tau_{l+1}^*)} + Z_m^{(l)} 1_{H_\alpha^{(l)}(L^{(m)}(T_l)) \geq 0} \right) \to \sup_{\alpha \in A},$$

where $Z_m^{(l)} := Z^{(l)}(L^{(m)}(T_l))$. If $\alpha_l^* \in A$ is a parameter where the supremum is attained, the family $H_{\alpha_l^*}^{(l)}, H_*^{(l+1)}, ..., H_*^{(k)}$ provides a backward extension of the exercise boundary to the exercise dates $T_l, ..., T_k$. In the ideal situation the parametric family $\{H_\alpha^{(l)}, \alpha \in A\}$ contains the optimal $H_*^{(l)}$ and then, naturally, $H_{\alpha_l^*}^{(l)} \to H_*^{(l)}$ (in a certain sense) as $M \to \infty$. In practice, however, it is usually not possible to identify such a family, and M can not be arbitrarily large. But, it may be possible to find a simple family which admits a close approximation of the optimal strategy, and then, for large enough M, the sequence $H_{\alpha_l^*}^{(l)}, H_*^{(l+1)}, ..., H_*^{(k)}$ may be considered to give a good approximation for the exercise boundary at the exercise dates $T_l, ..., T_k$. Next, we proceed to compute $H_{\alpha_{l-1}^*}^{(l-1)}$, down to $H_{\alpha_0^*}^{(0)}$ in the same way. The backward construction starts with $l = k - 1$. As a most important feature, *one and the same set of sample paths can be used for each recursion step*.

Bermudan Swaptions

As an example, we consider a Bermudan payer swaption with strike θ on the underlying swaps S_{T_j, T_n},[1] $0 \leq j \leq k$, with swap starting date T_j and fixed swap end T_n. The Bermudan discounted cash-flows at the exercise tenors are then given by

$$Z^{(j)} := \frac{B_{T_j, T_n}(S_{T_j, T_n}(T_j) - \theta)^+}{B(T_j)}, \quad 0 \leq j \leq k, \tag{5.10}$$

---

[1] For obvious reasons, we here choose a notation for swap rate and annuity numeraire which differs slightly from the respective notations in Chapters 1-4.

where B_{T_p,T_q} is the annuity numeraire; see (1.24) or (1.26). In the present notations, Andersen (1999) proposed the following strategies.

For parameters $\alpha := (\alpha^{(0)}, ..., \alpha^{(k-1)})^T \in R^{k-1}$, $0 \leq j < k$ (note that we set $H_*^{(k)} \equiv 0$), we may take

Exercise strategy I:

$$H_\alpha^{(j)}(L(T_j)) = B(T_j)Z^{(j)} - \alpha^{(j)};$$

Exercise strategy II:

$$H_\alpha^{(j)}(L(T_j)) = \min(B(T_j)Z^{(j)} - \alpha^{(j)}, \; Z^{(j)} - \max_{j<l\leq k} E^{\mathcal{F}^{(j)}} Z^{(l)});$$

Exercise strategy III:

$$H_\alpha^{(j)}(L(T_j)) = B(T_j)Z^{(j)} - \max_{j<l\leq k} B(T_j)E^{\mathcal{F}^{(j)}} Z^{(l)} - \alpha^{(j)}.$$

Remarkably, as we will see from our experimental results in Section 5.8.3 and also reported in Andersen (1999) and Andersen & Broadie (2001), the simple strategy I works pretty well in practice.

5.4 Iterative Construction of the Optimal Stopping Time

As an alternative to the backward construction of the exercise boundary in the previous section we now discuss the iterative procedure for approximating the Snell envelope of Kolodko & Schoenmakers (2004). We first present a procedure which improves upon a given exercise policy for the Bermudan problem (5.1), represented by a family of stopping times.

5.4.1 A One Step Improvement upon a Given Family of Stopping Times

With respect to the discrete filtration $\left(\mathcal{F}^{(j)}\right)_{0\leq j\leq k}$ we consider some given family of integer valued stopping indexes (τ_i), with the following properties,

$$i \leq \tau_i \leq k, \; \tau_k = k,$$
$$\tau_i > i \Rightarrow \tau_i = \tau_{i+1}, \qquad 0 \leq i < k, \tag{5.11}$$

and the process

$$Y^{(i)} := E^{\mathcal{F}^{(i)}} Z^{(\tau_i)}. \tag{5.12}$$

For example one could take

$$\tau_i := k \wedge \inf\{j \geq i : (\mathcal{T}_j, L(\mathcal{T}_j)) \in R\},$$

where R is a certain region in $\mathbb{R}^n$, or, as a more trivial example, the family $\tau_i \equiv i$. Generally, the process $(Y^{(i)})$ is a lower approximation of the Snell envelope process $(Y^{*(i)})$ due to the family of (sub-optimal) stopping times (τ_i).

Based on the family (τ_i) we are going to construct a new family $(\widehat{\tau}_i)$ satisfying (5.11), which induces a new approximation of the Snell envelope. We first introduce an intermediate process

$$\widetilde{Y}^{(i)} := \max_{p:\, i \leq p \leq k} E^{\mathcal{F}^{(i)}} Z^{(\tau_p)}. \tag{5.13}$$

Using $\widetilde{Y}^{(i)}$ as a new exercise criterion we define a next family of stopping indexes

$$\widehat{\tau}_i : = \inf\{j : i \leq j \leq k, \; \widetilde{Y}^{(j)} \leq Z^{(j)}\} \tag{5.14}$$
$$= \inf\{j : i \leq j \leq k, \; \max_{p:\, j \leq p \leq k} E^{\mathcal{F}^{(j)}} Z^{(\tau_p)} \leq Z^{(j)}\}, \quad 0 \leq i \leq k,$$

and consider the process

$$\widehat{Y}^{(i)} := E^{\mathcal{F}^{(i)}} Z^{\widehat{\tau}_i} \tag{5.15}$$

as a next approximation of the Snell envelope. Clearly, the family $(\widehat{\tau})$ satisfies the properties (5.11) as well. For instance, the trivial family $\tau_i \equiv i$ gives for $\widetilde{Y}$ the maximum of still alive Europeans and for $\widehat{Y}$ the Bermudan price due to the stopping rule "wait until the first exercise date when the cash-flow is at least equal to the maximum of still alive Europeans". See for a more detailed explanation of these examples Section 5.8.2. As another example, $\tau_i \equiv k$ gives for $\widetilde{Y}$ the European option process due to the last exercise date k and

$$\widehat{\tau}_i := \inf\{j : i \leq j \leq k, \; E^{\mathcal{F}^{(j)}} Z^{(k)} \leq Z^{(j)}\}, \quad 0 \leq i \leq k.$$

By the next crucial theorem, $(\widehat{Y}^{(i)})$ is generally an improvement of $(Y^{(i)})$.

THEOREM 5.1
(Kolodko & Schoenmakers (2004)) Let (τ_i) be a family of stopping times with the property (5.11) and $(Y^{(i)})$ be given by (5.12). Let the processes $(\widetilde{Y}^{(i)})$ and $(\widehat{Y}^{(i)})$ be defined by (5.13) and (5.15), respectively. Then, it holds

$$Y^{(i)} \leq \widetilde{Y}^{(i)} \leq \widehat{Y}^{(i)} \leq Y^{*(i)}, \quad 0 \leq i \leq k.$$

PROOF
The inequalities $Y^{(i)} \leq \widetilde{Y}^{(i)}$ and $\widehat{Y}^{(i)} \leq Y^{*(i)}$ are trivial. We only need to

show the middle inequality. We use induction in i. Due to the definition of $\widetilde{Y}$ and $\widehat{Y}$, we have $\widetilde{Y}^{(k)} = \widehat{Y}^{(k)} = Z^{(k)}$. Suppose that $\widetilde{Y}^{(i)} \leq \widehat{Y}^{(i)}$ for some i with $0 < i \leq k$. We will then show that $\widetilde{Y}^{(i-1)} \leq \widehat{Y}^{(i-1)}$. Let us write

$$\widehat{Y}^{(i-1)} = E^{\mathcal{F}^{(i-1)}} Z^{(\widehat{\tau}_{i-1})} = 1_{\widehat{\tau}_{i-1}=i-1} Z^{(i-1)} + 1_{\widehat{\tau}_{i-1}>i-1} E^{\mathcal{F}^{(i-1)}} E^{\mathcal{F}^{(i)}} Z^{(\widehat{\tau}_i)}$$
$$= 1_{\widehat{\tau}_{i-1}=i-1} Z^{(i-1)} + 1_{\widehat{\tau}_{i-1}>i-1} E^{\mathcal{F}^{(i-1)}} \widehat{Y}^{(i)}.$$

Then, by induction,

$$\widehat{Y}^{(i-1)} \geq 1_{\widehat{\tau}_{i-1}=i-1} Z^{(i-1)} + 1_{\widehat{\tau}_{i-1}>i-1} E^{\mathcal{F}^{(i-1)}} \widetilde{Y}^{(i)}$$
$$= 1_{\widehat{\tau}_{i-1}=i-1} Z^{(i-1)} + 1_{\widehat{\tau}_{i-1}>i-1} E^{\mathcal{F}^{(i-1)}} \max_{p:\, i\leq p\leq k} E^{\mathcal{F}^{(i)}} Z^{(\tau_p)}$$
$$\geq 1_{\widehat{\tau}_{i-1}=i-1} \widetilde{Y}^{(i-1)} + 1_{\widehat{\tau}_{i-1}>i-1} \max_{p:\, i\leq p\leq k} E^{\mathcal{F}^{(i-1)}} Z^{(\tau_p)}, \qquad (5.16)$$

since for $\widehat{\tau}_{i-1} = i-1$ we have $i-1 = \inf\{j : i-1 \leq j < k,\ \widetilde{Y}^{(j)} \leq Z^{(j)}\}$, and so

$$\widetilde{Y}^{(i-1)} = \max_{p:\, i-1\leq p\leq k} E^{\mathcal{F}^{(i-1)}} Z^{(\tau_p)} \leq Z^{(i-1)}.$$

We may write (5.16) as

$$\widehat{Y}^{(i-1)} \geq \widetilde{Y}^{(i-1)} + 1_{\widehat{\tau}_{i-1}>i-1}\Big(\max_{i\leq p\leq k} E^{\mathcal{F}^{(i-1)}} Z^{(\tau_p)} - \max_{i-1\leq p\leq k} E^{\mathcal{F}^{(i-1)}} Z^{(\tau_p)}\Big).$$

Thus, after showing that $\widehat{\tau}_{i-1} > i-1$ implies

$$E^{\mathcal{F}^{(i-1)}} Z^{(\tau_{i-1})} \leq \max_{p:\, i\leq p\leq k} E^{\mathcal{F}^{(i-1)}} Z^{(\tau_p)},$$

it follows that $\widehat{Y}^{(i-1)} \geq \widetilde{Y}^{(i-1)}$. It holds,

$$E^{\mathcal{F}^{(i-1)}} Z^{(\tau_{i-1})} = 1_{\tau_{i-1}=i-1} Z^{(i-1)} + 1_{\tau_{i-1}>i-1} E^{\mathcal{F}^{(i-1)}} Z^{(\tau_i)}$$
$$\leq 1_{\tau_{i-1}=i-1} Z^{(i-1)} + 1_{\tau_{i-1}>i-1} \max_{p:\, i\leq p\leq k} E^{\mathcal{F}^{(i-1)}} Z^{(\tau_p)}. \quad (5.17)$$

Then, on the set $\widehat{\tau}_{i-1} > i-1$ we have

$$Z^{(i-1)} < \max_{p:\, i-1\leq p\leq k} E^{\mathcal{F}^{(i-1)}} Z^{(\tau_p)},$$

so if $(\widehat{\tau}_{i-1} > i-1) \wedge (\tau_{i-1} = i-1)$ it follows that

$$Z^{(i-1)} < \max(Z^{(i-1)}, \max_{p:\, i\leq p\leq k} E^{\mathcal{F}^{(i-1)}} Z^{(\tau_p)}).$$

Hence, if $(\widehat{\tau}_{i-1} > i-1) \wedge (\tau_{i-1} = i-1)$,

$$Z^{(i-1)} < \max_{p:\, i\leq p\leq k} E^{\mathcal{F}^{(i-1)}} Z^{(\tau_p)}.$$

Thus, from (5.17) we have for $\widehat{\tau}_{i-1} > i - 1$,

$$E^{\mathcal{F}^{(i-1)}} Z^{(\tau_{i-1})} \leq 1_{\tau_{i-1}=i-1} \max_{p:\, i \leq p \leq k} E^{\mathcal{F}^{(i-1)}} Z^{(\tau_p)}$$

$$+ 1_{\tau_{i-1} > i-1} \max_{p:\, i \leq p \leq k} E^{\mathcal{F}^{(i-1)}} Z^{(\tau_p)} = \max_{p:\, i \leq p \leq k} E^{\mathcal{F}^{(i-1)}} Z^{(\tau_p)}.$$

□

5.4.2 Iterating to the Optimal Stopping Time and the Snell Envelope

Naturally, we may construct by induction via the procedure (5.14)-(5.15) a sequence of pairs

$$\left((\tau_i^{(m)})_{0 \leq i \leq k},\, (Y^m{}^{(i)})_{0 \leq i \leq k} \right)_{m=0,1,2,\ldots}$$

in the following way: Start with some family of stopping times $(\tau_i^{(0)})_{0 \leq i \leq k}$, which satisfies (5.11) and the additional requirement,

$$Y^0{}^{(i)} := E^{\mathcal{F}^{(i)}} Z^{(\tau_i^{(0)})} \geq Z^{(i)}, \quad 0 \leq i \leq k. \tag{5.18}$$

A canonical starting family is obtained, for example, by taking $\tau_i^{(0)} \equiv i$. Suppose that for $m \geq 0$ the pair $\left((\tau_i^{(m)}),\, (Y^m{}^{(i)}) \right)$ is constructed, where

$$Y^m{}^{(i)} := E^{\mathcal{F}^{(i)}} Z^{(\tau_i^{(m)})} \geq Z^{(i)}, \quad 0 \leq i \leq k,$$

and the stopping time family $(\tau_i^{(m)})$ satisfies (5.11). Then define

$$\tau_i^{(m+1)} := \inf\{j : i \leq j \leq k, \ \max_{p:\, j \leq p \leq k} E^{\mathcal{F}^{(j)}} Z^{(\tau_p^{(m)})} \leq Z^{(j)}\}$$

$$= \inf\{j : i \leq j \leq k, \ \widetilde{Y}^{m+1}{}^{(j)} \leq Z^{(j)}\}, \quad 0 \leq i \leq k, \tag{5.19}$$

with

$$\widetilde{Y}^{m+1}{}^{(i)} := \max_{p:\, i \leq p \leq k} E^{\mathcal{F}^{(i)}} Z^{(\tau_p^{(m)})}$$

being an intermediate dummy process and

$$Y^{m+1}{}^{(i)} := E^{\mathcal{F}^{(i)}} Z^{(\tau_i^{(m+1)})} \geq Z^{(i)}, \quad 0 \leq i \leq k.$$

Clearly, $\tau_i^{(m+1)}$ satisfies (5.11) as well, and due to Theorem 5.1 we have

$$Z^{(i)} \leq Y^0{}^{(i)} \leq Y^m{}^{(i)} \leq \widetilde{Y}^{m+1}{}^{(i)} \leq Y^{m+1}{}^{(i)} \leq Y^*{}^{(i)}, \ 0 \leq m < \infty, \ 0 \leq i \leq k. \tag{5.20}$$

By the following proposition, for each fixed i the sequence $(\tau_i^{(m)})_{m \geq 1}$ constructed above is non-decreasing in m and bounded by any optimal stopping time τ_i^*.

PROPOSITION 5.1

Let τ_i^* be an optimal stopping time. For each m: $1 \leq m < \infty$ and i: $0 \leq i \leq k$, we have

$$\tau_i^{(m)} \leq \tau_i^{(m+1)} \leq \tau_i^*.$$

PROOF   Suppose that $\tau_i^* < \tau_i^{(m)}$ for some $m \geq 1$ and some i with $0 \leq i \leq k$. Then, by (5.20) and the definition of $\tau_i^{(m)}$,

$$Y^*(\tau_i^*) \geq \widetilde{Y}^m(\tau_i^*) > Z^{(\tau_i^*)};$$

so τ_i^* is not optimal, hence a contradiction. Thus, the right inequality is proved. Next suppose $\tau_i^{(m+1)} < \tau_i^{(m)}$ for some $m \geq 1$ and some i with $0 \leq i \leq k$. Then, by the definition of $\tau_i^{(m)}$ we have

$$\widetilde{Y}^m(\tau_i^{(m+1)}) > Z^{(\tau_i^{(m+1)})}.$$

On the other hand, according to the definition of $\tau_i^{(m+1)}$, we have

$$\widetilde{Y}^{m+1}(\tau_i^{(m+1)}) \leq Z^{(\tau_i^{(m+1)})}.$$

So, we get $\widetilde{Y}^m(\tau_i^{(m+1)}) > \widetilde{Y}^{m+1}(\tau_i^{(m+1)})$, which contradicts (5.20). ⬚

Due to the inequality chain (5.20) and Proposition 5.1 we now may define a limit lower bound process Y^∞ and a limit family of stopping times (τ_i^∞) by

$$Y^\infty(i) := \text{(a.s.)} \lim_{m \uparrow \infty} \uparrow Y^m(i) \quad \text{and} \quad \tau_i^\infty := \text{(a.s.)} \lim_{m \uparrow \infty} \uparrow \tau_i^{(m)}, \quad 0 \leq i \leq k, \tag{5.21}$$

where the uparrows indicate that the respective sequences are non-decreasing. Further it is clear that the family (τ_i^∞) satisfies (5.11). Since $\tau_i^{(m)}$ is an integer valued random variable in the set $\{i, \dots, k\}$ for each m, we have with probability 1, $\tau_i^{(m)}(\omega) = \tau_i^\infty(\omega)$ for $m > N(\omega)$. Therefore, $Z^{(\tau_i^{(m)})} \to Z^{(\tau_i^\infty)}$ with probability 1, and so by dominated convergence we have

$$Y^\infty(i) = \text{(a.s.)} \lim_{m \uparrow \infty} \uparrow E^{\mathcal{F}(i)} Z^{(\tau_i^{(m)})} = E^{\mathcal{F}(i)} Z^{(\tau_i^\infty)}, \quad 0 \leq i \leq k.$$

We are now ready to present our main result.

THEOREM 5.2

(Kolodko & Schoenmakers (2004)) The constructed limit process Y^∞ in (5.21) coincides with the Snell envelope process Y^*, and (τ_i^∞) in (5.21) acts as a

family of optimal stopping times. We have

$$Y^*{(i)} = Y^{\infty}{(i)} = E^{\mathcal{F}^{(i)}} Z^{(\tau_i^{\infty})}, \quad 0 \leq i \leq k. \tag{5.22}$$

PROOF The Snell envelope $\left(Y^*{(i)}\right)$ is the smallest supermartingale which dominates the process Z (see, e.g., Shiryayev (1978), Elliot & Kopp, 1999). So, by (5.21) and the first inequality in (5.20) it is enough to show that the process $\left(Y^{\infty}{(i)}\right)$ is a supermartingale. Note, that for each i: $0 \leq i < k$,

$$
\begin{aligned}
E^{\mathcal{F}^{(i)}} Y^{\infty}{(i+1)} &= E^{\mathcal{F}^{(i)}} E^{\mathcal{F}^{(i+1)}} Z^{(\tau_{i+1}^{\infty})} \\
&= E^{\mathcal{F}^{(i)}} 1_{\tau_i^{\infty}=i} E^{\mathcal{F}^{(i+1)}} Z^{(\tau_{i+1}^{\infty})} + E^{\mathcal{F}^{(i)}} 1_{\tau_i^{\infty}>i} E^{\mathcal{F}^{(i+1)}} Z^{(\tau_{i+1}^{\infty})} \\
&= E^{\mathcal{F}^{(i)}} 1_{\tau_i^{\infty}=i} Z^{(\tau_{i+1}^{\infty})} + 1_{\tau_i^{\infty}>i} Y^{\infty}{(i)} \\
&= Y^{\infty}{(i)} + 1_{\tau_i^{\infty}=i}(E^{\mathcal{F}^{(i)}} Z^{(\tau_{i+1}^{\infty})} - Y^{\infty}{(i)}).
\end{aligned}
\tag{5.23}
$$

It is not difficult to see that

$$\tau_i^{\infty} = \inf\{j : i \leq j \leq k, \max_{p:\, j \leq p \leq k} E^{\mathcal{F}^{(j)}} Z^{(\tau_p^{\infty})} \leq Z^{(j)}\}, \quad 0 \leq i \leq k, \tag{5.24}$$

by letting $m \uparrow \infty$ in (5.19), the definition of $\tau_i^{(m)}$. Indeed, from (5.19) it follows that

$$
\begin{aligned}
\tau_i^{\infty} = \lim_{m \uparrow \infty} \tau_i^{(m)} &= \lim_{m \uparrow \infty} \inf\{j : i \leq j \leq k, \widetilde{Y}^m{(j)} \leq Z^{(j)}\} \\
&\leq \inf\{j : i \leq j \leq k, \widetilde{Y}^{\infty}{(j)} \leq Z^{(j)}\}, \quad 0 \leq i \leq k,
\end{aligned}
$$

since $\widetilde{Y}^m{(j)}$ converges to $\widetilde{Y}^{\infty}{(j)}$, non-decreasingly in m. Now suppose that

$$\tau_i^{\infty} < \inf\{j : i \leq j \leq k, \widetilde{Y}^{\infty}{(j)} \leq Z^{(j)}\}.$$

Then it follows that for m large enough

$$\widetilde{Y}^{\infty}{(l)} \geq \widetilde{Y}^m{(l)} > Z^{(l)}, \quad i \leq l \leq \tau_i^{\infty},$$

hence, $\tau_i^{(m)} > \tau_i^{\infty}$, a contradiction with (5.21).

Due to (5.24), for each j, $0 \leq j \leq k$, with $j < \tau_i^{\infty}$ we have

$$\max_{p:\, j \leq p \leq k} E^{\mathcal{F}^{(j)}} Z^{(\tau_p^{\infty})} > Z^{(j)},$$

and

$$\max_{p:\, \tau_i^{\infty} \leq p \leq k} E^{\mathcal{F}^{(\tau_i^{\infty})}} Z^{(\tau_p^{\infty})} \leq Z^{(\tau_i^{\infty})}.$$

Then, in particular,

$$1_{\tau_i^{\infty}=i} E^{\mathcal{F}^{(i)}} Z^{(\tau_{i+1}^{\infty})} \leq 1_{\tau_i^{\infty}=i} \max_{p:\, i \leq p \leq k} E^{\mathcal{F}^{(i)}} Z^{(\tau_p^{\infty})} \leq 1_{\tau_i^{\infty}=i} Z^{(i)}, \quad 0 \leq i < k.$$

Therefore, from (5.23) we obtain

$$E^{\mathcal{F}^{(i)}} Y^{\infty (i+1)} - Y^{\infty (i)} = 1_{\tau_i^\infty = i} (E^{\mathcal{F}^{(i)}} Z^{(\tau_{i+1}^\infty)} - Y^{\infty (i)}) \qquad (5.25)$$
$$\leq 1_{\tau_i^\infty = i} (Z^{(i)} - Y^{\infty (i)}) \leq 0, \quad 0 \leq i < k,$$

where the last inequality follows from (5.20). So the process $(Y^{\infty (i)})$ is a supermartingale. □

In this section the presented procedure for constructing the sequence $(Y^m)_{m \geq 0}$ has some flavour of Picard iteration. We may therefore expect rapid convergence in some sense. In this respect, Kolodko & Schoenmakers (2004) derived the following expression for the distance between two subsequent iterations,

$$Y^{m+1}(i) - Y^m(i) = E^{\mathcal{F}^{(i)}} \sum_{p=i}^{\tau_i^{(m+1)} - 1} 1_{\tau_p^{(m)} = p} (E^{\mathcal{F}^{(p)}} Y^{m (p+1)} - Y^{m (p)}) \geq 0. \quad (5.26)$$

The proof is given in Appendix A.3.2. Note that trivially $Y^{m (k)} = Z^{(k)}$ is constant for $m \geq 0$. Then, by using backward induction, the following funny result can be derived from (5.26); see Appendix A.3.3 and Bender & Schoenmakers (2004).

PROPOSITION 5.2
For $0 \leq i \leq k$ it holds

$$m \geq k - i \Longrightarrow Y^{m+1}(i) = Y^m(i) = Y^*(i) \quad a.s.$$

As a consequence, after k iterations the Snell envelope is constructed.

Note that for constructing the whole Snell envelope the backward dynamic program requires k iterations also. However, for a small m (e.g., $m = 2$), iteration Y^m (e.g., Y^2) provides an approximation of the *whole* Snell envelope, whereas m steps in the backward dynamic program give only the values $Y^{*(k-m)}, \ldots, Y^{*(k)}$.

Due to the definition of the stopping family $(\tau_i^{(m)})$, we have also the following result.

COROLLARY 5.1
Let $m \geq k - i$. Then,

$$\tau_i^{(m)} = \inf\{j : i \leq j \leq k, \ Y^*(j) \leq Z^{(j)}\},$$

i.e., $(\tau_i^{(m)})$ coincides with the first optimal stopping time for $m \geq k - i$.

5.4.3   On the Implementation of the Iterative Procedure

For the usual situation where the filtration $(\mathcal{F}^{(i)})$ is generated by a discrete Markov process (e.g., as in Section 5.3), the implementation of the proposed iterative procedure in Section 5.4.2 is straightforward and can be done in a generic way for a variety of (not necessarily financial) optimal stopping problems. Although such an implementation gives rise to nested Monte Carlo simulation, the degree of nesting can be chosen and, in particular, does not explode like in a Monte Carlo implementation of the backward dynamic program (see Section 5.2). Moreover, in Section 5.7 we will see that for Bermudan swaptions practically correct prices can be obtained already with $m = 2$ via a quadratic Monte Carlo simulation.

Let us consider a Markovian Libor process L (e.g., a market model) and assume that its dynamics is given in a numeraire B like in Remark 5.1. For example, B is the terminal bond. Consider a Bermudan Libor derivative for which the exercise dates coincide with the whole tenor structure, and the underlying European options can be priced (quasi-)analytically (a Bermudan swaption for example). Then, Algorithm 5.1 below sketches the iterative procedure in Section 5.4.2 for the case $m = 2$ and initial stopping family $\tau_i^{(0)} \equiv i$. Note that, on a fixed path ω, the set of exercise dates where a policy τ says "exercise" is characterised by

$$\{\eta : 0 \leq \eta \leq k, \tau_\eta(\omega) = \eta\}. \tag{5.27}$$

ALGORITHM 5.1

Simulate M trajectories $(L^{(m)}(T_j), \; j = 0, \ldots, n-1)$, $m = 1, \ldots, M$, starting at $L(0) \; (= L(T_0))$;

Along each trajectory (m) do the following:

   $i := 0;$

(A) Search the first exercise date $\geq i$ where the policy $\tau^{(1)}$ says "exercise"; hence search $\tau_i^{(1)}$ (e.g., in case of a Bermudan swaption we may use European swaption approximation formulas for this search). This date, say η, is a candidate for $\tau_0^{(2),(m)}$. To decide whether $\eta \approx \tau_0^{(2),(m)}$ or not we do the following:

   Simulate M_1 trajectories $(L^{(m,p)}(T_q), \; q = \eta, \ldots, n-1)$, $p = 1, \ldots, M_1$, with start value $L^{(m)}(T_\eta)$;

   Along each trajectory (m, p) search all exercise dates $\geq \eta$ where the policy $\tau^{(1)}$ says "exercise" (e.g. for the Bermudan swaption we may use European swaption approximation formulas again). From these dates we can detect easily via (5.27) (an

approximation of) the family
$(\tau_q^{(1),(m,p)},\ q \geq \eta)$ *along the path* (m,p);

Then, for $q = \eta, \ldots, k$ *compute*

$$Dummy[q] := \frac{1}{M_1} \sum_{p=1}^{M_1} Z^{(\tau_q^{(1),(m,p)})} \approx E^{L^{(m)}(T_\eta)} Z^{(\tau_q^{(1)})};$$

Next determine

$$Max\_Dummy := \max_{\eta \leq q \leq k} Dummy[q]$$

$$\approx \max_{\eta \leq q \leq k} E^{L^{(m)}(T_\eta)} Z^{(\tau_q^{(1)})};$$

Check whether $Z^{(\eta)} \geq Max\_Dummy$:

    If yes, set $\eta^{(m)} := \eta \approx \tau_0^{(2),(m)}$;

    If no, do $i := \eta + 1$; *if* $i < k$ *go to (A) and repeat;*

We so end with up an η *for which* $\eta^{(m)} := \eta \approx \tau_0^{(2),(m)}$;

Finally compute $\dfrac{1}{M} \displaystyle\sum_{m=1}^{M} Z^{(\eta^{(m)})} \approx E^{\mathcal{F}^{(0)}} Z^{(\tau_0^{(2)})}.$

The generalisation of Algorithm 5.1 to $m > 2$ is straightforward. Generally, provided that simple expectations $E^{\mathcal{F}^{(i)}} Z^{(j)}$ (Europeans) can be priced analytically, computation of Y^m requires about $O(N^m)$ simulations of the underlying process. For a person who is only interested in the optimal stopping time (for instance, the optimal exercise decision for the buyer of a Bermudan product), the stopping time $\tau^{(m)}$ can be decided at a cost of $O(N^{m-1})$ simulations.

REMARK 5.2 *(variance reduced Monte Carlo simulation of* Y^m*)* We can reduce the number of Monte Carlo simulations for Y^m by using the following variance reduced representation,

$$Y^{m(i)} = E^{\mathcal{F}^{(i)}} Z^{(\tau_i^{(m)})} = E^{\mathcal{F}^{(i)}} Z^{(\tau_i^{(m-1)})} + E^{\mathcal{F}^{(i)}} (Z^{(\tau_i^{(m)})} - Z^{(\tau_i^{(m-1)})}). \quad (5.28)$$

One can expect that $Z^{(\tau_i^{(m-1)})}$ and $Z^{(\tau_i^{(m)})}$ are strongly correlated and thus the variance of $(Z^{(\tau_i^{(m)})} - Z^{(\tau_i^{(m-1)})})$ will be less than the variance of $Z^{(\tau_i^{(m)})}$. So, the computation of $E^{\mathcal{F}^{(i)}} (Z^{(\tau_i^{(m)})} - Z^{(\tau_i^{(m-1)})})$ for a given accuracy can usually be done with less Monte Carlo simulations than needed for direct simulation of $E^{\mathcal{F}^{(i)}} Z^{(\tau_i^{(m)})}$. $\quad\square$

Finally, we note that the efficiency of the general iterative procedure for solving discrete optimal stopping problems proposed in this section can be improved in combination with several variance reduction techniques, for instance, by using (5.28) in combination with antithetic variables. Even quasi-Monte Carlo methods look promising in this respect. In any case, we expect that, when the producers of microprocessor chips keep "riding the exponential", computation of higher order iterations, so almost exact prices for Bermudan products, will become feasible in the near future.

5.5   Duality; From Tight Lower Bounds to Tight Upper Bounds

In fact, the procedures presented in Section 5.3 and Section 5.4 both provide suboptimal stopping rules for the Bermudan problem and consequently lower estimates for the Bermudan value. In this section we discuss the dual approach which enables the construction of tight upper bounds, from tight lower bounds, in some sense.

5.5.1   Dual Approach

The dual method for the optimal stopping problem goes back to Davis & Karatzas (1994), but is further developed in Rogers (2001) in the context of Monte Carlo simulation of continuous time American options, and independently in Haugh & Kogan (2001) for Bermudan style derivatives. The dual method for the Bermudan problem is based on the following observation. For any supermartingale $(S^{(j)})$ with $S^{(0)} = 0$ we have

$$
\begin{aligned}
Y^{*(0)} = \sup_{\tau \in \{0,\dots,k\}} E^{\mathcal{F}_0} Z^{(\tau)} &\leq \sup_{\tau \in \{0,\dots,k\}} E^{\mathcal{F}_0}(Z^{(\tau)} - S^{(\tau)}) \\
&\leq E^{\mathcal{F}_0} \max_{0 \leq j \leq k} (Z^{(j)} - S^{(j)});
\end{aligned} \tag{5.29}
$$

hence the right-hand side provides a (dual) upper bound for $Y^{*(0)}$.

Rogers (2001) and Haugh & Kogan (2001) show that the equality in (5.29) is attained at the martingale part of the Doob-Meyer decomposition of Y^*,

$$
M^{Y^*}(0) = 0; \quad M^{Y^*}(j) = \sum_{l=1}^{j} (Y^{*(l)} - E^{\mathcal{F}^{(l-1)}} Y^{*(l)}), \tag{5.30}
$$

and also at the shifted Snell envelope process

$$
S^{(j)} = Y^{*(j)} - Y^{*(0)}. \tag{5.31}
$$

The next lemma provides a somewhat more general class of supermartingales, which turn (5.29) into an equality. Moreover, we show that the equality holds almost sure.

LEMMA 5.1

Let S be the set of supermartingales S with $S^{(0)} = 0$. Let $S \in \mathcal{S}$ be such that $Z^{(j)} - Y^{*(0)} \leq S^{(j)}$, $1 \leq j \leq k$. Then,

$$Y^{*(0)} = \max_{0 \leq j \leq k} (Z^{(j)} - S^{(j)}) \quad a.s. \tag{5.32}$$

PROOF   From the assumptions it follows that $Z^{(j)} - Y^{*(0)} \leq S^{(j)}$, $0 \leq j \leq k$, then, using (5.29),

$$0 \leq E^{\mathcal{F}_0} \max_{0 \leq j \leq k} (Z^{(j)} - S^{(j)}) - Y^{*(0)} = E^{\mathcal{F}_0} \max_{0 \leq j \leq k} (Z^{(j)} - S^{(j)} - Y^{*(0)}) \leq 0.$$

Hence, we have $E^{\mathcal{F}_0} \max_{0 \leq j \leq k} (Z^{(j)} - S^{(j)} - Y^{*(0)}) = 0$ and $Z^{(j)} - S^{(j)} - Y^{*(0)} \leq 0$, which yields (5.32).   ∎

Note that both (5.30) and (5.31) satisfy the conditions of Lemma 5.1; for more examples see Kolodko & Schoenmakers (2004).

REMARK 5.3   Jamshidian (2003,2004) has presented a multiplicative analogue to the dual representation, both in continuous and discrete time. In the discrete setting we have analogue to (5.29),

$$Y^{*(0)} = \inf_{N \in \mathcal{M}} E^{\mathcal{F}_0} \max_{0 \leq j \leq k} Z^{(j)} \frac{N^{(k)}}{N^{(j)}}, \tag{5.33}$$

where $\mathcal{M}$ is the set of positive martingales $(N^{(j)})$, with $N^{(0)} = 1$. The derivation of the multiplicative representation (5.33) is similar to the additive case and, moreover, it is not difficult to show that the infimum in (5.33) is attained at the martingale part of the multiplicative Doob-Meyer decomposition of Y^*,

$$N^{Y^*(0)} = 1; \quad N^{Y^*(j)} = \prod_{l=1}^{j} \frac{Y^{*(l)}}{E^{\mathcal{F}^{(l-1)}} Y^{*(l)}}. \tag{5.34}$$

                                                                 ∎

5.5.2   Converging Upper Bounds from Lower Bounds

The duality representation in Section 5.5.1 provides a simple way to estimate the Snell envelope from above, by using an approximation of the Snell envelope denoted by $\overline{Y}$. Let $\overline{M}$ be the martingale part of the Doob-Meyer decomposition of $\overline{Y}$, satisfying

$$\overline{M}^{(0)} = 0;$$
$$\overline{M}^{(j)} = \overline{M}^{(j-1)} + \overline{Y}^{(j)} - E^{\mathcal{F}^{(j-1)}}\overline{Y}^{(j)}$$
$$= \sum_{l=i+1}^{j} \overline{Y}^{(l)} - \sum_{l=i+1}^{j} E^{\mathcal{F}^{(l-1)}}\overline{Y}^{(l)}, \ j = 1, \ldots, k.$$

Then, according to (5.29),

$$Y^{*(0)} \leq E^{\mathcal{F}_0} \max_{0 \leq j \leq k} (Z^{(j)} - \overline{M}^{(j)}) =: \overline{Y}^{(0)}_{up}. \tag{5.35}$$

If $\overline{Y}$ is a lower bound process which dominates the discounted cash-flows, then the gap between $\overline{Y}^{(0)}$ and $\overline{Y}^{(0)}_{up}$ depends, in some sense, on how far the lower bound process $\overline{Y}$ is away from being a supermartingale.

THEOREM 5.3
(Kolodko & Schoenmakers (2003,2004a)) Suppose that $Y^{*(j)} \geq \overline{Y}^{(j)} \geq Z^{(j)}$, $0 \leq j \leq k$. Then,

$$0 \leq \overline{Y}^{(0)}_{up} - \overline{Y}^{(0)} \leq E^{\mathcal{F}_0} \sum_{j=0}^{k-1} \max(E^{\mathcal{F}^{(j)}}\overline{Y}^{(j+1)} - \overline{Y}^{(j)}, 0).$$

PROOF   By definition (5.35), we have

$$\overline{Y}^{(0)}_{up} = E^{\mathcal{F}_0} \max_{0 \leq j \leq k} (Z^{(j)} - \sum_{l=1}^{j} \overline{Y}^{(l)} + \sum_{l=1}^{j} E^{\mathcal{F}^{(l-1)}}\overline{Y}^{(l)}) \tag{5.36}$$

$$= \overline{Y}^{(0)} + E^{\mathcal{F}_0} \max_{0 \leq j \leq k} (Z^{(j)} - \overline{Y}^{(j)} + \sum_{l=0}^{j-1} E^{\mathcal{F}^{(l)}}(\overline{Y}^{(l+1)} - \overline{Y}^{(l)}))$$

$$=: \overline{Y}^{(0)} + \Delta^{(0)}.$$

We thus have the following estimate,

$$\Delta^{(0)} = \overline{Y}^{(0)}_{up} - \overline{Y}^{(0)} \leq E^{\mathcal{F}_0} \max_{0 \leq j \leq k} \sum_{l=0}^{j-1} (E^{\mathcal{F}^{(l)}}\overline{Y}^{(l+1)} - \overline{Y}^{(l)})$$

$$\leq E^{\mathcal{F}_0} \max_{0 \leq j \leq k} \sum_{l=0}^{j-1} \max(E^{\mathcal{F}^{(l)}}\overline{Y}^{(l+1)} - \overline{Y}^{(l)}, 0)$$

$$\leq E^{\mathcal{F}_0} \sum_{j=0}^{k-1} \max(E^{\mathcal{F}^{(j)}}\overline{Y}^{(j+1)} - \overline{Y}^{(j)}, 0).$$

By the following Theorem 5.4 we can deduce a sequence of upper bound processes convergent to the Snell envelope from a sequence of lower bound processes which converges to the Snell envelope.

THEOREM 5.4
Let us consider a sequence $(Y^m)_{m \geq 0}$ of lower bound processes which dominate the discounted cash-flows, hence $Y^{m}{(j)} \leq Y^*{(j)}$ and $Y^{m}{(j)} \geq Z^{(j)}$, for $j = 0, \ldots, k$, $m \geq 0$, and which converge to the Snell envelope, i.e., $Y^*{(j)} = \lim_{m \to \infty} Y^{m}{(j)}$, for $j = 0, \ldots, k$. Then, analogue to (5.35) we can construct a sequence of upper bound processes by

$$Y_{up}^{m}{(i)} := E^{\mathcal{F}^{(i)}} \max_{i \leq j \leq k} \left(Z^{(j)} - \sum_{l=i+1}^{j} Y^{m}{(l)} + \sum_{l=i+1}^{j} E^{\mathcal{F}^{(l-1)}} Y^{m}{(l)} \right)$$

$$=: Y^{m}{(i)} + \Delta^{m}{(i)},$$

and the sequence $(Y_{up}^{m})_{m \geq 0}$ converges to the Snell envelope also.

PROOF   By Theorem 5.3 it follows that

$$0 \leq \Delta^{m}{(i)} \leq E^{\mathcal{F}^{(i)}} \sum_{j=i}^{k-1} \max \left(E^{\mathcal{F}^{(j)}} Y^{m}{(j+1)} - Y^{m}{(j)}, \, 0 \right).$$

By letting $m \uparrow \infty$ on the right-hand side, using dominated convergence, and the supermartingale property of Y^*, it follows that

$$(\text{a.s.}) \lim_{m \to \infty} \Delta^{m}{(i)} = 0, \quad 0 \leq i \leq k.$$

Hence,

$$(\text{a.s.}) \lim_{m \to \infty} Y_{up}^{m}{(i)} = (\text{a.s.}) \lim_{m \to \infty} Y^{m}{(i)} = Y^*{(i)}, \quad 0 \leq i \leq k.$$

Since we have constructed explicitly a convergent sequence of lower bound processes in Section 5.4, we may conclude this section with the following important corollary.

COROLLARY 5.2
The convergent sequence of lower bounds $(Y^m)_{m \geq 0}$, constructed in Section 5.4, induces via Theorem 5.4 a sequence of upper bounds $(Y_{up}^{m})_{m \geq 0}$, which converges to the Snell envelope as well. Moreover, due to Proposition 5.2, the upper bounds coincide with the Snell envelope after finitely many iteration steps as well.

5.6   Monte Carlo Simulation of Upper Bounds

Consider a Bermudan Libor derivative as introduced in Section 5.2 and suppose we have some approximative process $\widetilde{V}_t$ for the price of this Bermudan issued at time t. For instance, for some exercise strategy, i.e., a family of integer valued stopping times $\{\tau_t \in \{\kappa(t), ..., k\} : t \geq 0\}$, the process

$$\widetilde{V}_t := B(t) E^{\mathcal{F}_t} \frac{C_{\mathcal{T}_{\tau_t}}}{B(\mathcal{T}_{\tau_t})} = B(t) E^{\mathcal{F}_t} Z^{(\tau_t)} \qquad (5.37)$$

is a lower approximation, $\widetilde{V}_t \leq V_t$. However, here $\widetilde{V}_t$ doesn't need to be a strict lower bound, nor a strict upper bound. The discounted process $\widetilde{Y} := \widetilde{V}/B$ can be considered as a $\widetilde{V}$ associated approximation of the Snell envelope process Y^*. In the spirit of Section 5.5.2, definition (5.35), we obtain an upper bound for the Bermudan option at $t = 0$ via[2]

$$\frac{V_0}{B(0)} \leq \frac{V_0^{up}}{B(0)} =: E \sup_{1 \leq j \leq k} [Z^{(j)} - \sum_{i=1}^{j} \widetilde{Y}^{(i)} + \sum_{i=1}^{j} E^{\mathcal{F}^{(i-1)}} \widetilde{Y}^{(i)}]. \quad (5.38)$$

We now study Monte Carlo estimation of the target upper bound V_0^{up} in(5.38). For this we assume like in Section 5.3 that L is Markovian, and (without loss of generality) that the numeraire B is chosen as in Remark 5.1. For such a numeraire, $Y^{*(i)}$ is obviously a function of $L^{(i)}$ and we therefore assume that in (5.38) $\widetilde{Y}^{(i)}$ is a function of $L^{(i)}$ as well. Note that, for example, for the lower approximation (5.37) this is true. Then, due to Markovianity of L, in (5.38) $\mathcal{F}^{(i-1)}$ may be replaced by the σ-algebra generated by $L^{(i-1)} := L(\mathcal{T}_{i-1})$, for $i = 1, \ldots, k$. So, by general arguments there exist a (regular) conditional probability measure $P(L^{(j)}, \bullet)$, such that for any $\sigma\{L(s), s \geq \mathcal{T}_j\}$-measurable random variable Z,

$$[E^{\mathcal{F}^{(j)}} Z](\omega) = [E^{L^{(j)}} Z](\omega) =: \int P(L^{(j)}, d\widetilde{\omega}) Z(\widetilde{\omega}) \qquad a.s., \qquad j = 0, \ldots, k$$

(e.g., see Appendix A.1.4). We now consider for each j, $j = 1, ..., k$, a sequence of random variables $\left(\xi_i^{(j)} \right)_{i \in \mathbb{N}}$, where for $i \in \mathbb{N}$, $\xi_i^{(j)}$ are i.i.d. copies of $\widetilde{Y}^{(j)}$ under the conditional measure $P(L^{(j-1)}, \bullet)$, independent of the sigma-algebra $\sigma\{L^{(i)} : i = j, \ldots, k\}$. Hence,

$$E^{\mathcal{F}^{(j-1)}} \widetilde{Y}^{(j)} = \int P(L^{(j-1)}, d\widetilde{\omega}) \widetilde{Y}^{(j)}(\widetilde{\omega}) = \int P(L^{(j-1)}, d\widetilde{\omega}) \xi_i^{(j)}(\widetilde{\omega}), \qquad i \in \mathbb{N}.$$

---

[2]For minor technical reasons we henceforth assume without restriction that $t = \mathcal{T}_0 = 0$ is excluded from the set of exercise dates.

For a fixed but arbitrary $K \in \mathbb{N}$ we next consider a discrete process $\widetilde{M}^{(K)}$ defined by $\widetilde{M}_0^{(K)} = 0$ and, recursively,

$$\widetilde{M}_j^{(K)} := \widetilde{M}_{j-1}^{(K)} + \widetilde{Y}^{(j)} - \frac{1}{K}\sum_{i=1}^{K} \xi_i^{(j)}$$

$$= \sum_{q=1}^{j} \widetilde{Y}^{(q)} - \sum_{q=1}^{j} \frac{1}{K}\sum_{i=1}^{K} \xi_i^{(q)}, \qquad j = 1,\dots,k.$$

The process $\widetilde{M}^{(K)}$ is thus defined on an extended probability space $\Omega \times \prod$ with $\prod := \prod_{j=1}^{k} \mathbb{R}^K$. So a generic sample element in this space is $(\omega, (\xi^{(j)})_{1\leq j\leq k})$, with $\omega \in \Omega$ being a realisation of the process L and $\xi^{(j)} := (\xi_i^{(j)})_{i=1,\dots,K} \in \mathbb{R}^K$, for $j = 1,\dots,k$.

Clearly, $\widetilde{M}^{(K)}$ is a martingale w.r.t. the filtration $(\widetilde{\mathcal{F}}^{(j)})_{j=0,\dots,k}$, defined by $\widetilde{\mathcal{F}}^{(0)} := \mathcal{F}_0$ and $\widetilde{\mathcal{F}}^{(j)} := \sigma\{F\times H : \Omega \supset F \in \mathcal{F}^{(j)}, \prod \supset H \in \sigma\{\xi^{(1)},\dots,\xi^{(j)}\}\}$, for $j = 1,\dots,k$, and we observe that

$$E \sup_{1\leq j\leq k} [Z^{(j)} - \widetilde{M}_j^{(K)}] = EE^{\mathcal{F}^{(k)}} \sup_{1\leq j\leq k} [Z^{(j)} - \sum_{q=1}^{j}\widetilde{Y}^{(q)} + \sum_{q=1}^{j}\frac{1}{K}\sum_{i=1}^{K}\xi_i^{(q)}]$$

$$\geq E \sup_{1\leq j\leq k} [Z^{(j)} - \sum_{q=1}^{j}\widetilde{Y}^{(q)} + \sum_{q=1}^{j}\frac{1}{K}\sum_{i=1}^{K}E^{\mathcal{F}^{(k)}}\xi_i^{(q)}]$$

$$= E \sup_{1\leq j\leq k} [Z^{(j)} - \sum_{q=1}^{j}\widetilde{Y}^{(q)} + \sum_{q=1}^{j}\frac{1}{K}\sum_{i=1}^{K}E^{\mathcal{F}^{(q-1)}}\xi_i^{(q)}]$$

$$= E \sup_{1\leq j\leq k} [Z^{(j)} - \sum_{q=1}^{j}\widetilde{Y}^{(q)} + \sum_{q=1}^{j}E^{\mathcal{F}^{(q-1)}}\widetilde{Y}^{(q)}]$$

$$= \frac{V_0^{up}}{B(0)} \geq \frac{V_0}{B(0)},$$

where $E^{\mathcal{F}^{(k)}}\xi_i^{(q)} = E^{\mathcal{F}^{(q-1)}}\xi_i^{(q)}$ holds because $\xi_i^{(q)}$ is independent of $L^{(q)}, \dots, L^{(k)}$. Via the martingale $\widetilde{M}^{(K)}$ we have thus obtained a new upper bound

$$V_0^{up^{up},K} := B(0)E \sup_{1\leq j\leq k} [Z^{(j)} - \widetilde{M}_j^{(K)}], \tag{5.39}$$

which is larger than our target upper bound V_0^{up}.

The upper bound (5.39) is essentially used in the studies of Andersen & Broadie (2001) and Haugh & Kogan (2001). Usually, $V_0^{up^{up},K}$ in (5.39) will be already close to V_0^{up} for numbers K which are much smaller than the number of Monte Carlo trajectories needed for low variance estimation of the mathematical expectation (5.39). Nonetheless, due to the nested simulation required, Monte Carlo simulation of (5.39) is generally not very efficient in practice. In Section 5.8 we study (5.39) further and present new estimation procedures for V_0^{up} which are more efficient than (5.39).

5.7 Numerical Evaluation of Bermudan Swaptions by Different Methods

In this section we show some numerical results of the procedures presented in sections 5.3–5.6, applied to Bermudan swaptions. We consider a Libor market model where, as usual, we take for B the spot Libor rolling-over account as numeraire, hence the Libor dynamics (1.22), while taking into account Remark 5.1. The experiments are carried out with the volatility structure,

$$\gamma_i(t) = cg(T_i - t)e_i \quad \text{with} \quad g(s) = g_\infty + (1 - g_\infty + as)e^{-bs}$$

(see (2.25)), and with e_i being d-dimensional unit vectors, decomposing some input correlation matrix of rank d. For generating Libor models with different numbers of factors d, we take as basis a correlation structure of the form

$$\rho_{ij} = \exp(-\varphi|i - j|); \qquad i, j = 1, \ldots, n - 1, \tag{5.40}$$

which has full-rank for $\varphi > 0$. Then, for a particular choice of d, we deduce from ρ a rank-d correlation matrix ρ^d with decomposition $\rho_{ij}^d = e_i \cdot e_j$, $1 \leq i, j < n$, by principal component analysis (see Section 2.3.5). Of course, instead of (5.40) it is possible to use more general and economically more realistic correlation structures, for instance, the structure (2.46). As model parameters we take a flat 10% initial Libor curve over a 40 period quarterly tenor structure, and further

$$n = 41, \ \delta_i \equiv \delta = 0.25, \ c = 0.2, \ a = 1.5, \ b = 3.5, \ g_\infty = 0.5, \ \varphi = 0.0413. \tag{5.41}$$

The parameter values (5.41) are chosen such that the involved correlation structure and scalar volatilities can be regarded as typical for a Euro or GBP market. For a "practically exact" numerical integration of the SDE (1.22), we use the log-Euler scheme with $\Delta t = \delta/5$ (see, e.g., Chapter 6 and Kurbanmu-radov, Sabelfeld and Schoenmakers (2002)).

We now consider Bermudan swaptions for which the exercise dates coincide with the tenor structure between 1 yr. and 10 yr. According to the iterative method introduced in Section 5.4.2, we compute starting from $\tau_i^{(0)} \equiv i$, three successive lower bounds $Y^m(0)$: $Y^0(0)$, $Y^1(0)$ and $Y^2(0)$. Further, we compute lower estimations $Y_A^{(0)}$ obtained by Andersen's method outlined in Section 5.3,

TABLE 5.1:   Comparison of Andersen $Y_A^{(0)}$, Andersen & Broadie $Y_{up,A}^{(0)}$, and the iteration method $Y^{2(0)}$, OTM strike $\theta = 0.12$, base points

d	$Y^{0}(0)$ (SD)	$Y^{1}(0)$ (SD)	$Y^{2}(0)$ (SD)	$Y_A^{(0)}$ (SD)	$Y_{up,A}^{(0)}$ (SD)
1	10.2(0.0)	119.3(0.1)	131.0(0.9)	133.5(0.7)	135.4(0.1)
2	5.2(0.0)	114.5(0.1)	123.4(0.8)	119.7(0.7)	127.4(0.3)
10	3.0(0.0)	104.2(0.1)	110.5(0.6)	102.8(0.6)	113.6(0.3)
40	2.7(0.0)	101.4(0.1)	106.1(0.6)	98.8(0.5)	110.3(0.3)

and dual upper bounds $Y_{up,A}^{(0)}$, due to the lower bound process Y_A, via (5.39), as studied in Andersen & Broadie (2001). The process Y_A has the following form,

$$Y_A^{(i)} := E^{\mathcal{F}^{(i)}} Z^{(\tau_{A,i})}, \quad \text{with} \quad \tau_{A,i} := \inf \left\{ j \geq i : B(T_j)Z^{(j)} > \alpha^{(j)} \right\}, \quad (5.42)$$

where $Z^{(j)}$ are the discounted cash-flows according to (5.10), and the constants $\alpha^{(j)}$ are computed via backward induction; see Section 5.3 for details.

Later, in Section 5.8.3 we will see in the context of a study of efficient upper bound estimators that for 1-factor models the Andersen method due to strategy I gets very close to the Snell envelope. In fact, for one factor the relative distance between Y_A and by $Y_{up,A}$ does not exceed 1.5% in the examples of Section 5.8.3 (see for more examples Andersen & Broadie (2001)). However, in Section 5.8.3 we will see also that, when the number of factors gets larger, this distance increases from ITM to OTM strikes For OTM strikes and more than 2 factors this distance becomes even larger than 10% relative. For this reason we here consider OTM Bermudan swaptions for different numbers of factors, with an OTM strike $\theta = 0.12$. The simulation results for $Y_A^{(0)}$ and $Y_{up,A}^{(0)}$ given in Table 5.1 are taken over from Table 5.3 in Section 5.8.3.

For the iteration $Y^{2(0)}$, we apply the variance reduction technique (5.28). We use 5000000 Monte Carlo trajectories for $Y^{0}(0)$ and $Y^{1}(0)$ and 20 000 Monte Carlo trajectories (with 100 inner simulations) for computation of $Y^{2}(0)$ − $Y^{1}(0)$. As a consequence, the standard deviations of $Y^{2}(0)$ are roughly equal to the corresponding deviations due to the second term in (5.28), and are also comparable with the standard deviations of $Y_A^{(0)}$; see Table 5.1.

From Table 5.1 we see that in all cases, except for the 1-factor case, the secondly iterated lower bound $Y^{2}(0)$ is significantly higher than $Y_A^{(0)}$. Remarkably, for 10 and 40 factors already the first iteration $Y^{1}(0)$ is slightly higher than $Y_A^{(0)}$. In the 1-factor case, where Andersen's lower bound and corresponding dual upper bound are within 1.5% relative, $Y^{2}(0)$ is 1.5% relative below $Y_A^{(0)}$. Note that for more than 1 factor the computed lower bound $Y^{2}(0)$; hence the second iteration can be found more or less in the middle of $Y_A^{(0)}$ and $Y_{A,up}^{(0)}$.

REMARK 5.4   The construction of the new stopping time $\widehat{\tau}$ from τ via (5.14) provides a general method for improving any given stopping time τ with properties (5.11). So in principle we can improve Andersen's process Y_A by constructing $\widehat{Y}_A$ via (5.15). In this respect, we report that computations for two factor OTM cases yielded comparable values for $\widehat{Y}_A$ and $Y^2(0)$.   ☐

5.8   Efficient Monte Carlo Construction of Upper Bounds

As we have seen in Section 5.5, a good stopping rule can be used, on the one hand, to compute a tight lower bound of the Bermudan value, and on the other, to construct a tight upper bound via the duality approach. Unfortunately, however, whereas the lower bound can be computed by a straightforward Monte Carlo simulation, the upper bound construction via (5.39) requires in general a nested (quadratic) Monte Carlo simulation and is therefore not very efficient. In this section we present a method for a more efficient upper bound construction due to Kolodko & Schoenmakers (2003,2004a).

5.8.1   Alternative Estimators for the Target Upper Bound

We consider the set up of Section 5.6 again and, as a first step, we now construct a lower bound for the target upper bound V_0^{up} due to some approximative process $\widetilde{V}_t$. Consider an $(\mathcal{F}^{(k)})$-measurable random index $j_{\max}$ which satisfies

$$\sup_{1 \leq j \leq k} \left[Z^{(j)} - \sum_{q=1}^{j} \widetilde{Y}^{(q)} + \sum_{q=1}^{j} E^{\mathcal{F}^{(q-1)}} \widetilde{Y}^{(q)} \right] = Z^{(j_{\max})} - \sum_{q=1}^{j_{\max}} \widetilde{Y}^{(q)} + \sum_{q=1}^{j_{\max}} E^{\mathcal{F}^{(q-1)}} \widetilde{Y}^{(q)}.$$

Then, for any integer $K > 0$,

$$\frac{V_0^{up}}{B(0)} = E\left[Z^{(j_{\max})} - \sum_{q=1}^{j_{\max}} \widetilde{Y}^{(q)} + \sum_{q=1}^{j_{\max}} E^{\mathcal{F}^{(q-1)}} \widetilde{Y}^{(q)} \right]$$

$$= E\left[Z^{(j_{\max})} - \sum_{q=1}^{j_{\max}} \widetilde{Y}^{(q)} + \sum_{q=1}^{j_{\max}} \frac{1}{K} \sum_{i=1}^{K} E^{\mathcal{F}^{(q-1)}} \xi_i^{(q)} \right]$$

$$= E\left[Z^{(j_{\max})} - \sum_{q=1}^{j_{\max}} \widetilde{Y}^{(q)} + \sum_{q=1}^{j_{\max}} \frac{1}{K} \sum_{i=1}^{K} \xi_i^{(q)} \right],$$

because $E^{\mathcal{F}^{(k)}} \xi_i^{(q)} = E^{\mathcal{F}^{(q-1)}} \xi_i^{(q)} = E^{\mathcal{F}^{(q-1)}} \widetilde{Y}^{(q)}$. This brings us to the idea of localizing $j_{\max}$ for each particular simulation of the process L. To this aim, we carry out the following procedure. We consider on the extended probability space $\Omega \times \prod$ the random index $\widehat{\jmath}_{\max}$ which satisfies

$$Z^{(\widehat{\jmath}_{\max})} - \sum_{q=1}^{\widehat{\jmath}_{\max}} \widetilde{Y}^{(q)} + \sum_{q=1}^{\widehat{\jmath}_{\max}} \frac{1}{K} \sum_{i=1}^{K} \xi_i^{(q)} = \sup_{1 \le j \le k} \left[Z^{(j)} - \sum_{q=1}^{j} \widetilde{Y}^{(q)} + \sum_{q=1}^{j} \frac{1}{K} \sum_{i=1}^{K} \xi_i^{(q)} \right].$$

Next, we extend the probability space once again to $\Omega \times \prod \times \prod$ and simulate *independent copies* $\widehat{\xi}^{(j)} := (\widehat{\xi}_i^{(j)})_{i-1,\ldots,K} \in \mathbb{R}^K$, of $\xi^{(j)} \in \mathbb{R}^K$, for $j = 1, \ldots, k$. We then consider on $\Omega \times \prod \times \prod$ the random variable

$$Z^{(\widehat{\jmath}_{\max})} - \sum_{q=1}^{\widehat{\jmath}_{\max}} \widetilde{Y}^{(q)} + \sum_{q=1}^{\widehat{\jmath}_{\max}} \frac{1}{K} \sum_{i=1}^{K} \widehat{\xi}_i^{(q)}$$

with expectation

$$\frac{V_0^{up_{low},K}}{B(0)} := E\left[Z^{(\widehat{\jmath}_{\max})} - \sum_{q=1}^{\widehat{\jmath}_{\max}} \widetilde{Y}^{(q)} + \sum_{q=1}^{\widehat{\jmath}_{\max}} \frac{1}{K} \sum_{i=1}^{K} \widehat{\xi}_i^{(q)} \right]$$

$$= E\left[Z^{(\widehat{\jmath}_{\max})} - \sum_{q=1}^{\widehat{\jmath}_{\max}} \widetilde{Y}^{(q)} + \sum_{q=1}^{\widehat{\jmath}_{\max}} \frac{1}{K} \sum_{i=1}^{K} E^{\widetilde{\mathcal{F}}^{(k)}} \widehat{\xi}_i^{(q)} \right] \qquad (5.43)$$

$$= E\left[Z^{(\widehat{\jmath}_{\max})} - \sum_{q=1}^{\widehat{\jmath}_{\max}} \widetilde{Y}^{(q)} + \sum_{q=1}^{\widehat{\jmath}_{\max}} E^{\mathcal{F}^{(q-1)}} \widetilde{Y}^{(q)} \right]$$

$$\le E\left[Z^{(\widehat{\jmath}_{\max})} - \sum_{q=1}^{j_{\max}} \widetilde{Y}^{(q)} + \sum_{q=1}^{j_{\max}} E^{\mathcal{F}^{(q-1)}} \widetilde{Y}^{(q)} \right] = \frac{V_0^{up}}{B(0)},$$

where, most importantly, (5.43) holds while the $\widehat{\xi}^q$ are re-sampled *independent* of the determination of $\widehat{\jmath}_{\max}$ and then we have $E^{\widetilde{\mathcal{F}}^{(k)}} \widehat{\xi}_i^{(q)} = E^{\mathcal{F}^{(k)}} \widehat{\xi}_i^{(q)} = E^{\mathcal{F}^{(q-1)}} \widehat{\xi}_i^{(q)} = E^{\mathcal{F}^{(q-1)}} \widetilde{Y}^{(q)}$. As a result, $V_0^{up_{low},K}$ in (5.43) is a lower bound for V_0^{up}, for each K, $K = 1, 2, \ldots$ It is not difficult to show that for $K \to \infty$ both quantities converge to the target.

PROPOSITION 5.3

$$V_0^{up^{up},K} \downarrow V_0^{up} \quad \text{and} \quad V_0^{up_{low},K} \uparrow V_0^{up} \quad \text{for} \quad K \to \infty.$$

For a proof see Appendix A.3.4. We so come up with two different Monte Carlo estimators for the target upper bound V_0^{up}.

Lower estimate for V_0^{up}:

$$\widehat{V}_0^{up_{low},K,M} := \frac{B(0)}{M} \sum_{m=1}^{M} \left[Z^{(\widehat{j}_{\max}^{(m)})} - \sum_{q=1}^{\widehat{j}_{\max}^{(m)}} \widetilde{Y}^{(q;m)} + \sum_{q=1}^{\widehat{j}_{\max}^{(m)}} \frac{1}{K} \sum_{i=1}^{K} \xi_i^{(q;m)} \right] \quad (5.44)$$

Upper estimate for V_0^{up}:

$$\widehat{V}_0^{up^{up},K,M} := \frac{B(0)}{M} \sum_{m=1}^{M} \sup_{1 \le j \le k} \left[Z^{(j)} - \sum_{q=1}^{j} \widetilde{Y}^{(q;m)} + \sum_{q=1}^{j} \frac{1}{K} \sum_{i=1}^{K} \xi_i^{(q;m)} \right] \quad (5.45)$$

In (5.44), (5.45), $\widehat{j}_{\max}^{(m)}$ and $\widetilde{Y}^{(q;m)}$ denote the m-th independent sample of $\widehat{j}_{\max}$ and $\widetilde{Y}^{(q)}$, respectively. As a third alternative, motivated by Proposition 5.3, the estimators (5.44) and (5.45) can be combined into a convex family of new estimators,

Combined estimate for V_0^{up}:

$$\widehat{V}_0^{\alpha,K_u,K_l,M} := \alpha \widehat{V}_0^{up^{up},K_u,M} + (1-\alpha) \widehat{V}_0^{up_{low},K_l,M}, \quad (5.46)$$

for $0 < \alpha < 1$, and suitably chosen simulation numbers K_u, K_l, M.

In Section 5.8.3 we will demonstrate in practical examples that the combined estimator may have a much higher efficiency than either $\widehat{V}_0^{up^{up},K,M}$ or $\widehat{V}_0^{up_{low},K,M}$.

Heuristic motivation of the combined estimator

Let us suppose that, in view of Proposition 5.3, for some $\beta_u, \beta_l > 0$ the following expansions hold,

$$V_0^{up^{up},K} = V_0^{up} + \frac{c_u}{K^{\beta_u}} + o(\frac{1}{K^{\beta_u}}), \quad c_u > 0 \qquad \text{and}$$

$$V_0^{up_{low},K} = V_0^{up} - \frac{c_l}{K^{\beta_l}} + o(\frac{1}{K^{\beta_l}}), \quad c_l > 0. \quad (5.47)$$

Let α, $0 < \alpha < 1$, be such that $\alpha c_u - (1-\alpha)c_l = 0$ and let $\kappa_u, \kappa_l > 0$ be such that $\kappa_u \beta_u = \kappa_l \beta_l$. Consider for some integer K,

$$\widehat{V}_0^{\alpha,[K^{\kappa_u}],[K^{\kappa_l}]} := \alpha V_0^{up^{up},[K^{\kappa_u}]} + (1-\alpha) V_0^{up_{low},[K^{\kappa_l}]} = V_0^{up} + o(\frac{1}{K^{\kappa_u \beta_u}}), \quad (5.48)$$

with brackets denoting the Entier function. We then consider the complexity of the two estimators $\mathcal{U} := \widehat{V}_0^{up^{up},K,M}$ and $\mathcal{A} := \widehat{V}_0^{\alpha,[K^{\kappa_u}],[K^{\kappa_l}],M}$. As usual, the accuracy ε of an estimator $\widehat{s}$ for a target value p is defined via

$$\varepsilon^2 := E(\widehat{s} - p)^2 = Var(\widehat{s}) + (E\widehat{s} - p)^2$$

and so we may write by (5.47),(5.48),

$$\varepsilon_{\mathcal{U}}^2 := \frac{1}{M} Var(\widehat{V}_0^{up^{up},K,1}) + \frac{c_u^2}{K^{2\beta_u}} + o(\frac{1}{K^{2\beta_u}}),$$

$$\varepsilon_{\mathcal{A}}^2 := \frac{\alpha^2}{M} Var(\widehat{V}_0^{up^{up},[K^{\kappa_u}],1}) + \frac{(1-\alpha)^2}{M} Var(\widehat{V}_0^{up_{low},[K^{\kappa_l}],1}) + o(\frac{1}{K^{2\kappa_u\beta_u}}),$$

where the simulation of the up-up and up-low estimator is assumed to be done independently.

REMARK 5.5 In practice it is more efficient to localize $\widehat{j}_{\max}$ using the samples for $V_0^{up^{up},K}$. Then, the up-up and up-low estimator are dependent in general, so

$$\varepsilon_{\mathcal{A}}^2 := \frac{1}{M} Var\left(\alpha \widehat{V}_0^{up^{up},[K^{\kappa_u}],1} + (1-\alpha)\widehat{V}_0^{up_{low},[K^{\kappa_l}],1}\right) + o(\frac{1}{K^{2\kappa_u\beta_u}}).$$

☐

Since $Var(\widehat{V}_0^{up^{up},K,1})$ and $Var\left(\alpha \widehat{V}_0^{up^{up},[K^{\kappa_u}],1} + (1-\alpha)\widehat{V}_0^{up_{low},[K^{\kappa_l}],1}\right)$ are uniformly bounded in K, we can deduce in the spirit of Schoenmakers & Heemink (1997) and Duffy & Glyn (1995) an asymptotically optimal tradeoff between bias and statistical error of the estimators $\mathcal{U}$ and $\mathcal{A}$. In fact, their bias and statistical error should be of comparable magnitude. For the up-up estimator $\mathcal{U}$ we thus take $K \propto M^{1/(2\beta_u)}$ yielding $\varepsilon_{\mathcal{U}}^2 \propto M^{-1}$, with $\propto$ denoting asymptotic equivalence, and then for the required computational costs to achieve an accuracy ε we have

$$\text{Cost}_{\mathcal{U}}(\varepsilon) \propto MK \propto M^{1+\frac{1}{2\beta_u}} \propto \frac{1}{\varepsilon^{2+1/\beta_u}}.$$

For the combined estimator $\mathcal{A}$ we need to know a bit more about the bias term $o(K^{-\kappa_u\beta_u})$ in (5.48). Suppose we can identify a $\gamma > 1$, preferably as large as possible, such that this term may be represented as $O(K^{-\gamma\kappa_u\beta_u})$. Then, by choosing $K \propto M^{1/(2\kappa_u\beta_u\gamma)}$ we obtain in a similar way $\varepsilon_{\mathcal{A}}^2 = O(M^{-1})$ and

$$\text{Cost}_{\mathcal{A}}(\varepsilon) \propto MK^{\max(\kappa_u,\kappa_l)} \propto M^{1+\frac{\max(\kappa_u,\kappa_l)}{2\kappa_u\beta_u\gamma}} \propto \frac{1}{\varepsilon^{2+\max(1/\beta_u,1/\beta_l)/\gamma}}.$$

So, under the assumptions above,

$$\frac{\text{Cost}_{\mathcal{U}}(\varepsilon)}{\text{Cost}_{\mathcal{A}}(\varepsilon)} \longrightarrow \infty \quad \text{as} \quad \varepsilon \downarrow 0, \qquad \text{if} \quad \frac{\beta_u}{\beta_l} < \gamma. \qquad (5.49)$$

REMARK 5.6 We see that the complexity of the combined estimator $\mathcal{A}$ does only depend on the ratio $\kappa_l/\kappa_u = \beta_u/\beta_l$ and thus may take $\kappa_u =$

$\min(1, \beta_l/\beta_u)$ and $\kappa_l = \min(1, \beta_u/\beta_l)$, such that $O(K)$ is always the order of the number of inner simulations. □

The above analysis, which is built on some additional assumptions however, explains at a heuristic level why the combined estimator may be superior in several applications; see Section 5.8.3.

5.8.2   Two Canonical Approximative Processes

Here we address two canonical lower bound processes for the general Bermudan style derivative which arise from two canonical exercise strategies.

Maximum of still alive European options

Suppose the option holder has arrived at a certain exercise date T_j, $1 \le j \le k$, and looks at which remaining underlying European instrument has the largest value. More precisely, he considers the stopping index defined by

$$\widetilde{\tau}^{(j)} := \inf \left\{ m \ge j : E^{\mathcal{F}^{(j)}} Z^{(m)} = \max_{j \le p \le k} E^{\mathcal{F}^{(j)}} Z^{(p)} \right\}. \qquad (5.50)$$

This index is clearly $\mathcal{F}^{(j)}$-measurable and the option holder has the right to pin down his exercise policy at T_j for whatever reason, by deciding at T_j to exercise at $T_{\widetilde{\tau}^{(j)}}$. In fact, this is the same as selling the Bermudan at T_j as a European option with exercise date $T_{\widetilde{\tau}^{(j)}}$, thus receiving a cash amount of $\widetilde{Y}^{(j)} B(T_j)$, with

$$\widetilde{Y}^{(j)}_{\max} := \max_{j \le p \le k} E^{\mathcal{F}^{(j)}} Z^{(p)} = E^{\mathcal{F}^{(j)}} Z^{(\widetilde{\tau}^{(j)})} \le Y^{*(j)}. \qquad (5.51)$$

The process $\widetilde{Y}$ in (5.51) is a lower estimation of the Snell envelope Y since the policy (5.50) is suboptimal. For instance, because the optimal policy is not $\mathcal{F}^{(j)}$-measurable.

Exercise when cash-flow equals maximum of still alive European options

It is clear that exercising a Bermudan at a time where the cash-flow is below the maximum price of the remaining underlying European options is never optimal. This suggests an alternative exercise strategy defined by the following stopping time,

$$\widehat{\tau}^{(j)} := \inf \left\{ m \ge j : Z^{(m)} \ge \max_{m \le p \le k} E^{\mathcal{F}^{(m)}} Z^{(p)} \right\},$$

yielding a lower approximation of the Snell envelope,

$$\widehat{Y}^{(j)}_{\max} := E^{\mathcal{F}^{(j)}} Z^{(\widehat{\tau}^{(j)})} \le Y^{*(j)}. \qquad (5.52)$$

Due to Proposition 5.1 we have

$$\tilde{Y}_{\max} \leq \hat{Y}_{\max} \leq Y^*;$$

hence the exercise policy $\hat{\tau}$ is better than $\tilde{\tau}$. In fact, the process $\hat{Y}_{\max}$ corresponds to strategy II with $\alpha \equiv 0$ in the Andersen (1999) method; see Section 5.3.

5.8.3 Numerical Upper Bounds for Bermudan Swaptions

In this section we investigate numerically the bias of the upper bound estimators (5.44) and (5.45) for different Bermudan swaptions in the Libor market model (1.22). We take the same volatility structure, model parameters, and exercise dates as in Section 5.7, so a flat 10% initial yield curve, parameters (5.41), and three month exercise dates between 1 yr. and 10 yr. As a lower approximation of the Snell envelope process we consider the maximum of still alive swaption process $\tilde{Y}_{\max}$. Hence, (5.51), where the European option is now a European swaption with discounted cash-flows $Z^{(j)}$ according to (5.10). We further assume that expansion (5.47) holds true, hence,

$$V_0^{up^{up},K} - V_0^{up} = \frac{c_u}{K^{\beta_u}} + o(\frac{1}{K^{\beta_u}}),$$

$$V_0^{up} - V_0^{up_{low},K} = \frac{c_l}{K^{\beta_l}} + o(\frac{1}{K^{\beta_l}}), \qquad \beta_u, \beta_l, c_u, c_l > 0, \qquad (5.53)$$

and aim to identify the parameters β_u, β_l, c_u, c_l in particular cases.

We compute $V_0^{up_{low},K}$ and $V_0^{up^{up},K}$ by estimators (5.44) and (5.45), respectively, with $K = 2^2, 2^3, \ldots, 2^7$ and $M = 30000$, for the examples in Table 5.2. For $M = 30000$ the standard deviations of both estimators are less than 1.5% relative, for all considered K. For $K = 128$, the relative distance between $\hat{V}_0^{up_{low},K,30000}$ and $\hat{V}_0^{up^{up},K,30000}$ turns out to be within 1.5%, hence the relative standard deviation of both estimators. So we conclude that within a relative accuracy of 1.5% in this sense, both estimators $\hat{V}_0^{up_{low},128,30000}$ and $\hat{V}_0^{up^{up},128,30000}$ give a good approximation of the target upper bound V_0^{up}. Therefore, we take their average $\hat{V}_0^{1/2,128,128,30000}$ as an approximation of V_0^{up}.

With regard to (5.53) we next determine the coefficients $\beta_u, \beta_l, c_u, c_l$ by linear regression, i.e., we carry out the following minimizations,

$$RMS_u^{rel} = \qquad\qquad (5.54)$$

$$\sqrt{\sum_{i=2}^{6} \left(\frac{\log(\hat{V}_0^{up^{up},2^i,30000} - \hat{V}_0^{1/2,128,128,30000}) - (\log c_u - \beta_u \log 2^i)}{\log(\hat{V}_0^{up^{up},2^i,30000} - \hat{V}_0^{1/2,128,128,30000})} \right)^2}$$

$$\longrightarrow \min_{\beta_u, c_u}, \quad \text{and}$$

$$RMS_l^{rel} = \tag{5.55}$$

$$\sqrt{\sum_{i=2}^{6}\left(\frac{\log(\widehat{V}_0^{1/2,128,128,30000} - \widehat{V}_0^{up_{low},2^i,30000}) - (\log c_l - \beta_l \log 2^i)}{\log(\widehat{V}_0^{1/2,128,128,30000} - \widehat{V}_0^{up_{low},2^i,30000})}\right)^2}$$

$$\longrightarrow \min_{\beta_l,c_l},$$

by straightforward differentiating. Note that in the linear regressions (5.54) and (5.55) we exclude terms due to $i = 7$. This is because for $i = 7$ the denominators in (5.54) and (5.55) are basically zero within the considered accuracy. The values of β_u, c_u, β_l, c_l, obtained for different types of swaptions and different number of factors d are given in Table 5.2. We also show in Table 5.2 the "optimal" $\alpha = c_l/(c_u + c_l)$ and ratios $\beta_l/\beta_u = \kappa_u/\kappa_l$.

Discussion of Numerical Results

According to Table 5.2, the function $\log(V_0^{up^{up},K} - V_0^{up})$ and $\log(V_0^{up_{low},K} - V_0^{up})$ can be approximated rather close by $\log c_u - \beta_u \log K$ and $\log c_l - \beta_l \log K$, respectively, within errors which do not exceed 3.0%. Hence plotting $\log K \to \log(V_0^{up^{up},K} - V_0^{up})$ and $\log K \to \log(V_0^{up_{low},K} - V_0^{up})$ gives approximately straight lines. See Figure 5.1 and 5.2 for $d = 40$ (full factor model) and out-of-the-money swaptions with strike $\theta = 12\%$. The values of β_u turn out to be roughly equal to one, whereas β_l seem to be significantly smaller than one over all. It would be interesting to study this issue in further detail. Then, it is remarkable that the optimal value of α for different strikes and number of factors does not vary too much. The same applies for β_u and β_l and we therefore consider for all examples the combined upper bound estimator

$$\widehat{V}_0^{0.4,[K^{0.87}],K,M} = 0.4\widehat{V}_0^{up^{up},[K^{0.87}],M} + 0.6\widehat{V}_0^{up_{low},K,M}, \tag{5.56}$$

where $\alpha = 0.4$ is roughly the average value in Table 5.2, and $\kappa_u = 0.87$ and $\kappa_l = 1$ are based on the average of β_l/β_u and taking into account Remark 5.6. In Figure 5.2 we show for a particular example, strike $\theta = 0.12$ (OTM) and $d = 40$, a plot of the estimator (5.56) together with $\widehat{V}_0^{up^{up},[K^{0.87}],30000}$ and $\widehat{V}_0^{up_{low},K,30000}$ for different values of K. Note that even for any K the bias of the combined estimator is negligible within the given accuracy in this example. Later (in Table 5.3) we will see that the bias of the estimator (5.56) for the particular choices of $\alpha, \kappa_u, \kappa_l$, is negligible also for all other examples in Table 5.2, when $K \geq 4$.

We now compare the combined estimator (5.56) with the up-up estimator (5.45) for different strikes and different number of factors d. We consider $\widehat{V}_0^{0.4,4,5,90000}$ and $\widehat{V}_0^{up^{up},100,30000}$, where the respective choices of K and M are determined by experiment, such that both the estimations and the (absolute) standard deviations of the estimators are close for different strikes and different number of factors. The results are given in Table 5.3, columns 5,6.

It is easily seen that the combined estimator $\widehat{V}_0^{0.4,4,5,90000}$ is almost 4 times faster than the up-up estimator $\widehat{V}_0^{up^{up},100,30000}$.

REMARK 5.7   In general, depending on the quality of the Snell-envelope approximation, higher accuracies for dual upper bound estimations may be required and then the efficiency gain of the combined up-low estimator (5.56) with respect to up-up estimator (5.45) can become tremendous in view of (5.49). ▯

Now we are going to compare the up-up estimations $V_0^{up^{up},K}$, based on the maximum of still alive swaption process (5.51), with up-up estimations due to an approximative lower bound process $\widetilde{Y}_A$, obtained via a particular exercise boundary, constructed by the Andersen (1999) method (see Section 5.3). The process $\widetilde{Y}_A$ is given in (5.42), where the sequence of constants (α_m) is pre-computed by using Andersen strategy I. We here denote the corresponding upper bound, which is studied in Andersen & Broadie (2001), as $V_{0,AB}^{up^{up},K}$.

We compute $\widehat{V}_{0,AB}^{up^{up},100,10000}$ for different strikes and number of factors, and the results are given in Table 5.3, column 4. As we can see, the values of $\widehat{V}_{0,AB}^{up^{up},100,10000}$ and $\widehat{V}_0^{up^{up},100,30000}$ are rather close. In fact, except for the ATM strikes in the one and two factor model, the differences do not exceed 1% relative. For a full factor model and a particular OTM strike we also compare the estimators $\widehat{V}_{0,AB}^{up^{up},K,10000}$ and $\widehat{V}_0^{up^{up},K,30000}$ for different numbers of inner simulations, $K = 1,\ldots,100$, and conclude that both estimators coincide within the considered accuracy; see Figure 5.2.

In Table 5.3, column 3, we give lower bounds of Bermudan prices $B^*(0)\widetilde{Y}_A^{(0)}$, due to the stopping time $\tau_A^{(0)}$ (see (5.42)). We see that in case of a one factor model the distance between the lower and upper bound of the Bermudan swaption price is rather close for OTM, ATM as well as for ITM strikes. This observation is consistent with the results reported in Andersen & Broadie (2001). For more than one factor however, the gap between lower and upper bound appears to increase. In particular, as we see in Table 5.3, when the number of factors is larger than 1, the gap increases from ITM to OTM strikes and gets even larger than 10% relative for OTM strikes. Apparently, for more factors the exercise region is more complicated in the sense that the exercise region can not be approximated properly by strategy I.

In Table 5.4 we list the required computational time[3] of the up-up upper bound estimators due to Andersen & Broadie and the process given by the maximum of still alive swaptions. We used the Euler scheme with time steps $\Delta t = \delta$ and for practical relevance, we required an accuracy of 1% relative. The cost of pre-computation of the exercise strategy is not taken into account in Table 5.4. In fact, this cost is small compared with the cost of the upper

---

[3]The simulations are run on a 1 GHz processor

estimators. It appears that for ATM and OTM strikes the upper bound method due to $\widetilde{Y}_{\max}$ is faster than the upper bound method due to $\widetilde{Y}_A$. The main reason is the fact that in the algorithm for simulating $\widetilde{Y}_A$ one needs to construct a Libor trajectory starting at T_j until the exercise condition is fulfilled (for a description of this algorithm see also Andersen & Broadie, (2001)).

Regarding the rather high computation times in Table 5.4, it is clear that an efficiency gain of about a factor 4 (or maybe more), due to application of the combined upper bound estimator, is desirable in practice. Moreover, for a particular Bermudan product we recommend the following procedure.

Carry out a pre-computation of the optimal β_u, β_l and α for the given structure, based on up-up and up-low estimations with lower accuracy. Next, take the number K of inner simulations as small as possible and then choose the number M of outer simulations according to the accuracy required.

For example see Figure 5.3, where the involved parameters β_u, β_l, and α are optimal for the example under consideration and where K can be taken equal to one in fact. We conclude this section with two further remarks.

REMARK 5.8 Naturally, the numerical analysis based on the maximum of still alive swaption process (5.51) could also be done for the process (5.52). This process is in fact consistent with Andersen strategy II, $\alpha_m \equiv 0$ (Section 5.3). So, on the one hand, this process is dominated from above by a lower bound process due to strategy II with an optimized (α_m). On the other hand, however, as Andersen reports and we found out also, strategy II with optimized (α_m) performs not substantially better than strategy I with optimized (α_m). Therefore, it is expected that the dual upper bound due to process (5.52) will be more or less comparable with the upper bound due to $\widetilde{Y}_A^{(0)}$, which in turn is comparable with the upper bound due to (5.51) for a more than one factor model. Moreover, it is easily seen that the computation of the dual upper bound by the process (5.52) will be more costly. ▯

REMARK 5.9 It would be interesting to consider the computational aspects of Jamshidian's (2003,2004) multiplicative dual method (5.33) (5.34) also. In particular, since Jamshidian's dual may also require nested simulation, it would be nice to have a similar combined estimator for reducing the number of inner simulations. Further, efficient upper bounds for so called Israeli options, Bermudans, which are cancelable from the issuer's side, would be interesting (see Kühn & Kyprianou (2003)). ▯

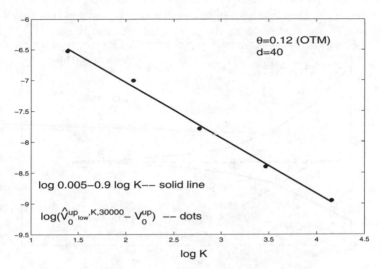

FIGURE 5.1: Coefficients in (5.53) for the up-low estimator via logarithmic regression: $\log(\widehat{V}_0^{up_{low},K,30000} - V_0^{up})$ and $\log c_l - \beta_l \log K$ for β_l and c_l minimizing (5.55)

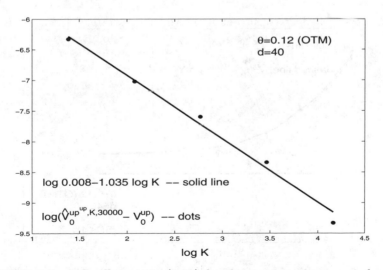

FIGURE 5.2: Coefficients in (5.53) for the up-up estimator via logarithmic regression: $\log(\widehat{V}_0^{up^{up},K,30000} - V_0^{up})$ and $\log c_u - \beta_u \log K$ for β_u and c_u minimizing (5.54)

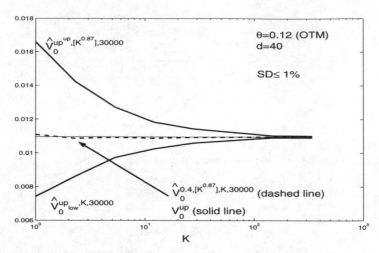

FIGURE 5.3:   Different Bermudan upper bound estimators due to max of still alive swaptions $\widetilde{Y}_{\max}$

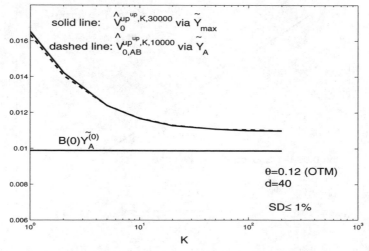

FIGURE 5.4:   "up^{up}" Bermudan upper bound estimators due to $\widetilde{Y}_{\max}$ and $\widetilde{Y}_A$

TABLE 5.2: Coefficients of (5.47) for different swaptions and numbers of factors

θ	d	β_u	c_u	RMS_u^{rel}	β_l	c_l	RMS_l^{rel}	β_l/β_u	α
	1	1.021	0.038	0.016	0.850	0.023	0.026	0.83	0.381
0.08	2	0.991	0.032	0.016	0.862	0.022	0.019	0.87	0.404
(ITM)	10	0.940	0.026	0.003	0.893	0.020	0.023	0.95	0.436
	1	0.970	0.025	0.021	0.746	0.013	0.015	0.77	0.335
0.10	2	0.872	0.020	0.009	0.840	0.014	0.029	0.96	0.417
(ATM)	10	0.968	0.021	0.016	0.717	0.010	0.020	0.74	0.317
	1	0.988	0.099	0.015	0.801	0.006	0.017	0.81	0.363
0.12	2	0.946	0.008	0.007	0.872	0.006	0.016	0.93	0.442
(OTM)	10	0.930	0.007	0.009	0.896	0.006	0.013	0.96	0.460
	40	1.035	0.008	0.029	0.900	0.005	0.019	0.87	0.405

TABLE 5.3: Upper bounds for different strikes, factors and approximative processes (values in base points)

θ	d	$B^*(0)\widetilde{Y}_A^{(0)}$ (SD)	$\widehat{V}_{0,AB}^{up^{up},100,10000}$ (SD)	$\widehat{V}_0^{up^{up},100,30000}$ (SD)	$\widehat{V}_0^{0.4,4,5,90000}$ (SD)
	1	1116.2(1.6)	1121.4(0.1)	1128.8(0.3)	1124.8(0.4)
0.08	2	1103.2(1.4)	1117.6(0.4)	1121.1(0.3)	1117.6(0.3)
(ITM)	10	1097.1(1.3)	1111.0(0.4)	1113.7(0.3)	1109.5(0.3)
	40	1093.2(1.3)	1106.9(0.4)	1110.1(0.3)	1106.9(0.3)
	1	403.3(1.2)	408.3(0.1)	416.5(0.5)	416.1(0.5)
0.10	2	372.6(1.1)	394.0(0.4)	397.3(0.5)	397.7(0.4)
(ATM)	10	347.4(1.0)	373.6(0.5)	375.8(0.4)	375.3(0.4)
	40	341.6(1.0)	367.5(0.5)	368.5(0.4)	369.8(0.4)
	1	133.5(0.7)	135.4(0.1)	136.3(0.4)	135.5(0.3)
0.12	2	119.7(0.7)	127.4(0.3)	127.5(0.3)	126.5(0.3)
(OTM)	10	102.8(0.6)	113.6(0.3)	114.5(0.3)	113.2(0.3)
	40	98.8(0.5)	110.3(0.3)	109.6(0.3)	108.6(0.3)

TABLE 5.4:   Computation times for
upper bounds (in sec.) of the upper
estimators in Table 5.3

θ	d	$\widehat{V}_{0,AB}^{up^{up},100,1000}$	$\widehat{V}_{0}^{up^{up},100,3000}$
	1	59	115
0.08	2	69	134
(ITM)	10	183	241
	40	468	603
	1	166	113
0.10	2	213	134
(ATM)	10	510	239
	40	1467	598
	1	229	115
0.12	2	299	145
(OTM)	10	718	263
	40	2076	625

5.9   Multiple Callable Structures

We conclude this chapter with an optional description of Bermudan structures, where the investor has in general more than one exercise right. A typical example is the chooser flex cap which can be regarded as a standard cap with the restriction that the option holder may call only a certain number of caplets during the life time of the cap. In this section we discuss a method of Bender & Schoenmakers (2004), which is a generalisation of the iterative procedure for standard Bermudans (Kolodko & Schoenmakers (2004), Section 5.4) to multiple callable products.

5.9.1   The Multiple Stopping Problem

Consider as in Section 5.2 a sequence $(Z(i): i = 0, 1, \ldots, k)$ of cash-flows, adapted to some filtration $(\mathcal{F}^{(i)} : 0 \leq i \leq k)$. We now think of an investor who has the right to exercise a cash-flow D_0 times under the additional constraint that he has to wait a minimal number of exercise dates $d_0 \in \mathbb{N}$, called *refracting index*, between exercising two rights. The addition of this constraint avoids trivialities, as otherwise the investor would exercise all rights at the same time. For Libor products usually $d_0 = 1$. The investor's problem is now to maximize his expected gain by exercising optimally.

We now formalize the multiple stopping problem. Due to the multiple setting we need additional indexes and therefore we now choose for a slightly different notation of stopping times, cash-flows etc. This change will maintain the readability without causing any confusion with the notations of the previous sections. Let us define $\mathcal{S}_i(D_0, d_0)$ as the set of $\mathcal{F}^{(i)}$ stopping vectors $(\tau_1(i), \ldots, \tau_{D_0}(i))$ such that $i \leq \tau_1(i)$ and, for all $2 \leq j \leq D_0$, $\tau_{j-1}(i) + d_0 \leq \tau_j(i)$. For convenience we use the convention $Z(i) = 0$ for $i \geq k+1$. The multiple stopping problem can now be stated as follows: Find for $0 \leq i \leq k$ a family of stopping vectors $\tau^*(i) \in \mathcal{S}_i(D_0, d_0)$, such that

$$E^{\mathcal{F}^{(i)}} \sum_{j=1}^{D_0} Z(\tau_j^*(i)) = \sup_{\tau \in \mathcal{S}_i(D_0, d_0)} E^{\mathcal{F}^{(i)}} \sum_{j=1}^{D_0} Z(\tau_j) =: Y_{D_0}^*(i).$$

The process $Y_{D_0}^*$ is called the *Snell envelope* of Z under D_0 exercise rights. For the simple stopping problem we usually write $Y^*(i) = Y_1^*(i)$.

The multiple stopping problem can be reduced to D_0 nested stopping problems with one exercise right in the following way. Consider a sequence of processes $(X_D)_{0 \leq D \leq D_0}$ with $X_0 \equiv 0$ and for $1 \leq D \leq D_0$, X_D being the Snell envelope of the cash-flows $Z(i) + E^{\mathcal{F}^{(i)}} X_{D-1}(i + d_0)$ under one exercise right. In particular, $X_1 = Y_1^* = Y^*$ is the Snell envelope of Z. We next define for $1 \leq D \leq D_0$,

$$\sigma_D^*(i) = \inf\{i \leq j \leq k : Z(j) + E^{\mathcal{F}^{(j)}} X_{D-1}(j + d_0) \geq X_D(j)\} \qquad (5.57)$$

as an optimal stopping family for the simple stopping problem with Snell envelope X_D. It is straightforward to show by induction over D, that for $1 \leq D \leq D_0$,

$$Y_D^*(i) = X_D(i), \qquad (5.58)$$

and that an optimal stopping vector $(\tau_{d,D})_{1 \leq d \leq D}$ for the multiple stopping problem with D exercise rights and cash-flow Z is given by

$$\tau_{1,D}^*(i) = \sigma_D^*(i)$$
$$\tau_{d+1,D}^*(i) = \tau_{d,D-1}^*(\sigma_D^*(i) + d_0) \quad 1 \leq d \leq D - 1. \qquad (5.59)$$

The above reduction is intuitively clear and basically says that the investor has to choose the first stopping time of the stopping vector in the following way: Decide, at time j, whether it is better to take the cash-flow $Z(j)$ and enter a new contract with $D - 1$ exercise rights starting at $j + d_0$, or to keep the D exercise rights. Then, after entering the stopping problem with $D - 1$ exercise rights, he proceeds to behave optimally. By this reduction, any algorithm for single optimal stopping problems can, in principle, be applied iteratively to the multiple stopping problem. For example, Carmona & Touzi (2004) suggest to apply backward dynamic programming iteratively to the D stopping problems. However, this leads to even higher nestings of conditional expectations than the dynamic programming approach for the single stopping problem. As an alternative, we here propose to generalize the policy iteration procedure for the simple Bermudan problem in Section 5.4 to an algorithm which simultaneously improves the Snell envelope for D exercise rights, where $D = 1, \ldots, D_0$.

5.9.2 Iterative Algorithm for Multiple Bermudan Products

We now explain how the procedure in Section 5.4 can be generalized to the multiple stopping problem, hence the pricing of multiple callable products. Consider a family of Bermudans with D exercise rights for $1 \leq D \leq D_0$ and refracting index d_0. Suppose we are given a policy $\sigma_D^{(0)}(i)$, $1 \leq D \leq D_0$, $0 \leq i \leq k$, as an initial guess for an optimal stopping family (5.57). We then may iterate this policy by induction in the following way. Let for $m \geq 0$,

$$\sigma_D^{(m)}(i), \quad 0 \leq i \leq k, \quad 1 \leq D \leq D_0,$$

be the m-th improvement upon the originally given family $(\sigma_D^{(0)}(i))_{1 \leq D \leq D_0}$. We may interpret $\sigma_D^{(m)}(i)$ as the exercise date when the investor exercises (according to the m-th improvement) the first of his D rights provided he has not exercised prior to time with index i. Then

$$\sigma_D^{(m+1)}(i), \quad 0 \leq i \leq k, \quad 1 \leq D \leq D_0,$$

is constructed as follows. First we define a family of stopping vectors $\tau_{1,D}^{(m)}$ by

$$\tau_{1,D}^{(m)}(i) = \sigma_D^{(m)}(i) \tag{5.60}$$

$$\tau_{d+1,D}^{(m)}(i) = \tau_{d,D-1}^{(m)}\left(\sigma_D^{(m)}(i) + d_0\right), \quad 1 \leq d \leq D-1,\ 2 \leq D \leq D_0,\ 0 \leq i \leq k,$$

where we set $\sigma_D^{(m)}(i) = i$ for $i \geq k+1$, $1 \leq D \leq D_0$ and recall the convention $Z(i) = 0$ for $i \geq k+1$. According to the m-th improvement, $\tau_{d,D}^{(m)}(i)$ can be interpreted as the exercise date, when the investor exercises the d-th of his D exercise rights, provided he has not exercised his first right prior to time i. The m-th approximation of the Snell envelope with D exercise rights due to the m-th improvement is given by

$$Y_D^{(m)}(i) = E^{\mathcal{F}^{(i)}} \sum_{d=1}^{D} Z(\tau_{d,D}^{(m)}(i)).$$

For constructing a next approximation of the Snell envelope we introduce an intermediate process by

$$\tilde{Y}_D^{(m+1)}(i) = \max_{p \geq i} E^{\mathcal{F}^{(i)}} \sum_{d=1}^{D} Z(\tau_{d,D}^{(m)}(p))$$

and use this as an exercise criterion to obtain a new stopping family,

$$\sigma_D^{(m+1)}(i) = \inf\left\{j \geq i;\ Z(j) + E^{\mathcal{F}^{(j)}} Y_{D-1}^{(m)}(j + d_0) \geq \tilde{Y}_D^{(m+1)}(j)\right\},$$

with the convention $Y_0^{(m)}(i) \equiv 0$. Then, via (5.60) we derive a stopping family $\tau_{1,D}^{(m+1)}$ and the $m+1$ approximation of the Snell envelope is given by

$$Y_D^{(m+1)}(i) = E^{\mathcal{F}^{(i)}} \sum_{d=1}^{D} Z(\tau_{d,D}^{(m+1)}(i)).$$

We now prove the following generalisation of Theorem 5.1.

THEOREM 5.5

(Bender & Schoenmakers (2004)) Suppose $\sigma_D^{(0)}(i)$ satisfies property (5.11) for all $1 \leq D \leq D_0$. Then, for all $m \geq 0$, $1 \leq D \leq D_0$, and $0 \leq i \leq k$,

$$Y_D^{(m+1)}(i) \geq Y_D^{(m)}(i).$$

Moreover, for $m \geq Dk - i$,

$$Y_D^{(m)}(i) = Y_D^*(i),$$

where $Y_D^*(i)$ denotes the Snell envelope of Z under D exercise rights at date i.

PROOF   We prove the theorem by induction over D. For $D = 1$ the algorithm coincides with the procedure described in Section 5.4 for which the results holds. For the step from $D - 1$ to D we start with some preliminary considerations. Define a family of auxiliary processes by

$$Z_D^{(m+1)}(i) = Z(i) + E^{\mathcal{F}^{(i)}} Y_{D-1}^{(m)}(i + d_0).$$

Then, for $p \geq j$,

$$E^{\mathcal{F}^{(j)}} Z_D^{(m+1)}(\sigma_D^{(m)}(p)) = E^{\mathcal{F}^{(j)}} \left[Z(\sigma_D^{(m)}(p)) + \sum_{d=1}^{D-1} Z(\tau_{d,D-1}^{(m)}(\sigma_D^{(m)}(p) + d_0)) \right]$$

$$= E^{\mathcal{F}^{(j)}} \left[Z(\tau_{1,D}^{(m)}(p)) + \sum_{d=1}^{D-1} Z(\tau_{d+1,D}^{(m)}(p)) \right]$$

$$= E^{\mathcal{F}^{(j)}} \left[\sum_{d=1}^{D} Z(\tau_{d,D}^{(m)}(p)) \right],$$

making use of the definition of $\tau_{d+1,D}^{(m)}(p)$. In particular,

$$Y_D^{(m)}(i) = E^{\mathcal{F}^{(i)}} Z_D^{(m+1)}(\sigma_D^{(m)}(i)) \tag{5.61}$$

$$\tilde{Y}_D^{(m+1)}(j) = \max_{p \geq j} E^{\mathcal{F}^{(j)}} Z_D^{(m+1)}(\sigma_D^{(m)}(p)). \tag{5.62}$$

Thus, the definition of $\sigma_D^{(m+1)}(i)$ can be rewritten as

$$\sigma_D^{(m+1)}(i) = \inf\left\{j \geq i;\ Z_D^{(m+1)}(j) \geq \max_{p \geq j} E^{\mathcal{F}^{(j)}}\, Z_D^{(m+1)}(\sigma_D^{(m)}(p))\right\}. \quad (5.63)$$

By (5.61)–(5.63) the step from $\sigma_D^{(m)}(i)$ to $\sigma_D^{(m+1)}(i)$ is a one step improvement as described in Section 5.4. Thus,

$$Y_D^{(m)}(i) \leq E^{\mathcal{F}^{(i)}}\, Z_D^{(m+1)}(\sigma_D^{(m+1)}(i)).$$

Now, by the induction hypothesis,

$$Z_D^{(m+1)}(i) = Z(i) + E^{\mathcal{F}^{(i)}}\, Y_{D-1}^{(m)}(i+d_0) \leq Z(i) + E^{\mathcal{F}^{(i)}}\, Y_{D-1}^{(m+1)}(i+d_0) = Z_D^{(m+2)}(i).$$

Hence, we may conclude from (5.61),

$$E^{\mathcal{F}^{(i)}}\, Z_D^{(m+1)}(\sigma_D^{(m+1)}(i)) \leq E^{\mathcal{F}^{(i)}}\, Z_D^{(m+2)}(\sigma_D^{(m+1)}(i)) = Y_D^{(m+1)}(i).$$

Therefore, $Y_D^m(i)$ is increasing in m.

Now fix $0 \leq i_0 \leq k$ and $m_0 \geq Dk - i_0$. By the induction hypothesis,

$$Y_{D-1}^{(m)}(i) = Y_{D-1}^*(i),$$

for all $m \geq (D-1)k - i$. In particular,

$$Z_D^{(m+1)}(i) = Z(i) + E^{\mathcal{F}^{(i)}}\, Y_{D-1}^{(m)}(i+d_0) = Z(i) + E^{\mathcal{F}^{(i)}}\, Y_{D-1}^*(i+d_0)$$

for all $0 \leq i \leq k$ and $m \geq (D-1)k$. This means that from step $(D-1)k$ we have an iteration procedure as in the case of a single exercise right, but with the cash-flow $Z(i)$ replaced by $Z(i) + E^{\mathcal{F}^{(i)}}\, Y_{D-1}^*(i+d_0)$. The time i value of such an iteration does not change after $k - i$ new improvements by Proposition 5.2, but coincides with the Snell envelope at that time. Hence, for $m_0 \geq Dk - i_0 = (D-1)k + k - i_0$, $Y_D^{(m_0)}(i_0)$ coincides with the time i_0 value of the Snell envelope of $Z(i) + E^{\mathcal{F}^{(i)}}\, Y_{D-1}^*(i+d_0)$ with one exercise right, which, in turn, equals $Y_D^*(i_0)$ by (5.58). □

We also have an analogue to Corollary 5.1.

COROLLARY 5.3
Let $m \geq Dk - i$. Then,

$$\sigma_D^{(m)}(i) = \inf\left\{j : i \leq j \leq k,\ Y_D^*(j) \leq Z(j) + E^{\mathcal{F}^{(j)}}\, Y_{D-1}^*(j+d_0)\right\}.$$

In particular, the stopping times $(\tau_{d,D}^{(m)}(i) : 1 \leq d \leq D)$ are optimal, when $m \geq Dk - i$.

PROOF   The representation of $\sigma_D^{(m)}(i)$ follows from Corollary 5.1 taking the second part of the proof of Theorem 5.5 into account. The optimality of $\tau_{d,D}^{(m)}(i)$ is then a simple consequence of (5.58)–(5.59).   ∎

Chapter 6

Pricing Long Dated Products via Libor Approximations

6.1 Introduction

In this chapter we study lognormal approximations for Libor market models, proposed in Kurbanmuradov, Sabelfeld and Schoenmakers (2002), where special attention is paid to their simulation by direct methods and lognormal random fields. Generally, the main advantage of log-normal approximations is that their distributions can be simulated fast since, in contrast to the usual numerical solution of the Libor SDE, the approximations can be simulated directly at any future point in time. For instance, the lognormal approximation proposed in Brace, Gatarek and Musiela (1997) can be simulated effectively by a Gaussian random field of log-Libors. As a result, since in general valuation of Libor derivatives comes down to computation of expected values of functions of Libors, an important family of Libor instruments consisting of long dated exotic products can be valuated faster by using lognormal approximations. As such the proposed approximations provide valuable alternatives to the Euler method, in particular for long dated instruments.

After introducing different lognormal Libor approximations we carry out a path-wise comparison with the "exact" SDE solution obtained by the Euler scheme using sufficiently small time steps. Also we test approximations obtained via numerical solution of the SDE by the Euler method, using larger time steps. It turns out that, for typical volatilities observed in practice, improved versions of the Brace, Gatarek and Musiela (1997) approximation appear to have excellent path-wise accuracy. Although this accuracy can be achieved also by Euler stepping the SDE using larger time steps as we will see, it follows from a comparative cost analysis that, particularly for long maturity options, the Euler method is more time consuming than the lognormal approximation. We illustrate this at some example Libor options with 12 yr. maturity.

In Section 6.2 we derive different lognormal approximations for a Libor market model in the terminal bond measure given by SDE (1.16) and study their accuracy. In particular, the lognormal approximations are subjected to a mutual path-wise comparison with approximations obtained by Euler stepping the SDE. A ranking between the different approximations is thus obtained. Then, in Section 6.3, we construct and investigate direct simulation techniques for the approximations derived in Section 6.2. The efficiency of the lognormal approximations with respect to Euler stepping the SDE is analysed in Section 6.4. Based on this analysis we propose an "optimal" simulation program in which different approximations may be combined. In Section 6.5 we consider the valuation of swaptions and trigger swaps for different approximations and compare the results. Also we discuss briefly the callable reverse floater, an exotic instrument for which the proposed simulation method is extremely efficient.

Throughout this chapter we work with the Libor market model in the terminal measure (1.16). It will be clear, however, how to convert the presented Libor approximations to a Libor SDE in another measure, and that generally the same conclusions as in Section 6.6 can be made. For the experiments in this chapter we used a model with time independent scalar volatilities and time independent instantaneous correlations due to a correlation generating sequence of the form (2.48),

$$b_i = \exp[\beta(i-1)^\alpha], \quad \beta > 0, \ 0 < \alpha < 1, \tag{6.1}$$

$$\rho_{ij}(t) \equiv \frac{\min(b_i, b_j)}{\max(b_i, b_j)} = \exp[-\beta|(i-1)^\alpha - (j-1)^\alpha|], \qquad 1 \le i, j < n.$$

This structure is more simple than (2.46) for instance, but it has full rank and is of ratio form (2.29) which satisfies (2.30). So, the structure (6.1) has realistic features yet. In particular, the condition $0 < \alpha < 1$ implies that instantaneous correlations $\rho(dL_i, dL_{i+p})$ are increasing in i for fixed p, which is very natural indeed.

6.2 Different Lognormal Approximations

Let us recall the Libor SDE (1.16) in the terminal measure P_n,

$$dL_i = - \sum_{j=i+1}^{n-1} \frac{\delta_j L_i L_j}{1 + \delta_j L_j} \gamma_i^\top \gamma_j dt + L_i \gamma_i^\top dW^{(n)}, \quad t_0 \le t \le T_i, \tag{6.2}$$

where $0 \le t_0 \le T_i$ and $1 \le i < n$. It is convenient to deal with (6.2) in the following integrated form,

$$\ln \frac{L_i(t)}{L_i(t_0)} = - \int_{t_0}^{t} \sum_{j=i+1}^{n-1} \frac{\delta_j L_j}{1 + \delta_j L_j} |\gamma_i||\gamma_j|\rho_{ij} \, ds - \frac{1}{2} \int_{t_0}^{t} |\gamma_i|^2 ds + \int_{t_0}^{t} \gamma_i^\top dW^{(n)},$$

$$\tag{6.3}$$

where $\rho_{ij} = \gamma_i^\top \gamma_j / |\gamma_i||\gamma_j|$. In practice, we may define the vectors $\gamma_i/|\gamma_i|$ through the matrix (ρ_{ij}) by applying a Cholesky decomposition.

Note that only the first term in the right hand side of (6.3) is generally non-Gaussian. Let us consider the contribution of the non-Gaussian term where we assume for simplicity that the functions γ_i are constants. We introduce the notations: $\rho_i = \sum_{j=i+1}^{n-1} |\rho_{ij}|$, $\rho = \max_i \rho_i$, $\delta = \max_i \delta_i$, and $\gamma = \max_i |\gamma_i|$.

Let us denote by $\tilde{L}$ the maximum value of the L_i, i.e., $\tilde{L} = \max_i \sup_{t_0 \leq t \leq T_i} L_i(t)$.

Then, we may write (6.3) as

$$\ln \frac{L_i(t)}{L_i(t_0)} = \varepsilon_i - \frac{1}{2}|\gamma_i|^2(t - t_0) + |\gamma_i|\sqrt{t - t_0}\, Z_i(t),$$

where $Z_i(t)$ is a standard normally distributed random variable and ε_i can be estimated by $|\varepsilon_i| \leq (t - t_0)\delta\tilde{L}\gamma^2\rho_i$. So, by neglecting ε_i we cause in L_i only a small *relative* error of order of ε_i when

$$|\varepsilon_i| \leq (t - t_0)\delta\tilde{L}\gamma^2\rho_i \leq (t - t_0)\delta\tilde{L}\gamma^2\rho << 1. \tag{6.4}$$

Note that for typical values, e.g., $\delta = 0.25$, $\gamma = 0.15$, $\tilde{L} = 0.07$, $t - t_0 = 10$, this relative error is about $0.4\rho\,\%$. However, dependent on ρ and the length of the tenor structure this error can become rather large in practice.

The approximation by neglecting the non-Gaussian terms ε_i in (6.3) will be called (0)-approximation to (6.2). In this approximation, forward Libor rates satisfy

$$dL_i^{(0)} = L_i^{(0)} \gamma_i^\top dW^{(n)}, \tag{6.5}$$

and are given by the explicit solution

$$L_i^{(0)}(t) = L_i(t_0) \exp \left\{ -\frac{1}{2}\int_{t_0}^{t} |\gamma_i|^2(s)ds + \int_{t_0}^{t} \gamma_i(s)^\top dW^{(n)}(s) \right\}. \tag{6.6}$$

In Figure 6.1 and Figure 6.2 we show for illustration some samples of $L_i(t)$ and $L_i^{(0)}(t)$. For these figures, scalar volatilities are taken to be relatively high, $|\gamma_1| = \ldots = |\gamma_{n-1}| = 0.4$, with $n = 31$, in order to amplify effects. The correlation matrix ρ_{ij} is given by the two parametric structure (6.1), where $\alpha = 0.8$, $\beta = 0.1$ in Figure 6.1, and $\alpha = 0.8$, $\beta = 0.3$ in Figure 6.2. Further, $t_0 = 0$ and the tenor structure is given by $T_i = i\delta$ with $\delta = 0.25$, $i = 1, \ldots, 31$. The initial L values are taken to be $L_i(0) = 0.061$. The "true" process $L_i(t)$ is simulated by an Euler scheme with time discretization step $\delta/10$.

From the trajectories in Figure 6.1 and Figure 6.2 we see that on an initial time interval, the trajectory $L_{10}^{(0)}$ approximates the trajectory L_{10} very well. For increasing time, however, the discrepancy increases. Also, in accordance with (6.4), for increasing ρ in (6.4), respectively decreasing β in (6.1), this

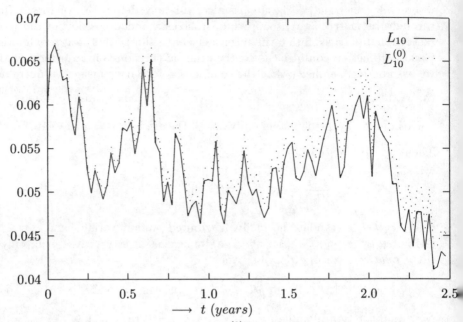

FIGURE 6.1:   Sample of $L_{10}(t)$ and $L_{10}^{(0)}(t)$, $\alpha = 0.8$ and $\beta = 0.1$

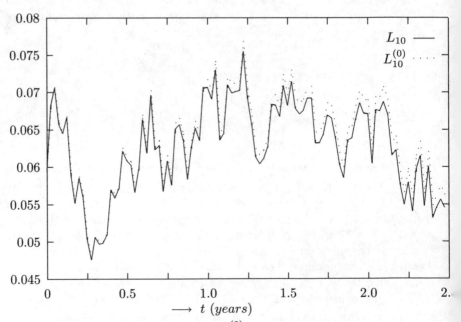

FIGURE 6.2:   Sample of $L_{10}(t)$ and $L_{10}^{(0)}(t)$, $\alpha = 0.8$ and $\beta = 0.3$

discrepancy increases. These effects are illustrated by Figure 6.1 and Figure 6.2 and their comparison. So, on the one hand, we see from the plots in Figures 6.1 and 6.2 that, in accordance with (6.4), the (0)-approximation performs well for small times. On the other hand, we see from (6.4) that the (0)-approximation produces good results also for large i, since ρ_i decreases with i (in particular, ρ_{n-1} vanishes). More details about the (0) and other approximations are presented in Tables 6.2– 6.5.

In Figure 6.3 we show a sample of the bond price $B_{31}(T_i)$ and its (0)-approximation $B_{31}^{(0)}(T_i)$, $i = 0, \ldots, 31$, for the same model parameters as used for Figure 6.1. The bond prices are computed from the Libor process, via the relationship

$$B_{31}(T_i) = \prod_{j=i}^{30}(1 + \delta L_j(T_i))^{-1}, \tag{6.7}$$

and a similar one for the (0)-approximations, where for the "true" bond prices we have used "true" Libors, simulated by an Euler scheme with time discretization step $\delta/10$.

In contrast to the results presented in Figures 6.1 and 6.2, the maximum discrepancy in Figure 6.3 occurs around the middle of the time interval $(0, T_{31})$. The reason is clear from (6.7). Indeed, when i is either close to zero or when it is close to 30, where the drift terms become small, the approximations $L_j^{(0)}(T_i)$, $j = i, \ldots, 30$ are close to $L_j(T_i)$ and so $B_{31}^{(0)}(T_i)$ is close to $B_{31}(T_i)$. Note that, exactly, $B_{31}^{(0)}(30 \times 0.25) = B_{31}(30 \times 0.25)$ because L_{30} and $L_{30}^{(0)}$ coincide in the terminal measure given by $n = 31$.

6.2.1 More Lognormal Approximations

It is of interest to consider more refined approximations to L and in particular to look for lognormal approximations improving $L^{(0)}$. So, in fact, we need to find a deterministic or normal approximation to the sum term in the integral representation (6.3). Therefore, for each j we approximate the process

$$Z_j(t) := \frac{\delta_j L_j}{1 + \delta_j L_j} \tag{6.8}$$

in the following way. Let the function f be defined as $f(x) := x/(1+x)$ and $Z_j = f(\delta_j L_j)$. Hence the process $Z := [Z_1, \ldots, Z_{n-1}]$ satisfies the SDE system

$$dZ_j = f'(\delta_j L_j)\delta_j L_j \gamma_j^\top dW^{(n)} + \frac{1}{2}f''(\delta_j L_j)\,(\delta_j L_j|\gamma_j|)^2\,dt$$

$$- \sum_{k=j+1}^{n-1} \frac{\delta_j \delta_k L_j L_k\,\gamma_j^\top \gamma_k}{1 + \delta_k L_k}\,f'(\delta_j L_j)dt \tag{6.9}$$

$$=: a_j(Z,t)dt + b_j(Z,t)^\top dW^{(n)}(t), \quad 1 \le j < n,$$

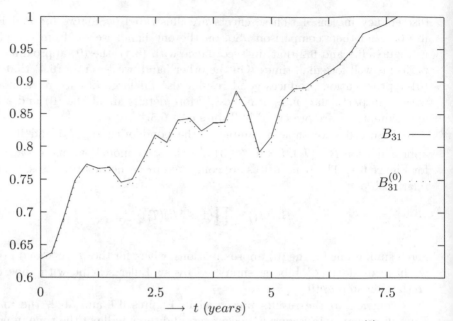

FIGURE 6.3:   Sample of $B_{31}(t)$ and its (0)-approximation $B_{10}^{(0)}(t)$, same parameters as for Figure 6.1

with initial conditions $Z_j(t_0) = f(\delta_j L_j(t_0))$. Note that by solving (6.8) for the $L_j's$ and substituting the result in (6.9) it is immediately clear how the coefficients a_j and b_j should be defined explicitly.

The Picard-Lindelof-0 and Picard-Lindelof-1 iteration for the solution of this SDE are

$$Z_j^{(0)}(t) := Z_j(t_0) = \frac{\delta_j L_j(t_0)}{1 + \delta_j L_j(t_0)} \qquad \text{and}$$

$$Z_j^{(1)}(t) := Z_j(t_0) + \int_{t_0}^{t} [a_j(Z^{(0)}(t_0), s)ds + b_j(Z^{(0)}(t_0), s)^\top dW^{(n)}(s)]$$

$$= f(\delta_j L_j(t_0)) + \frac{1}{2}f''(\delta_j L_j(t_0))\delta_j^2 L_j^2(t_0) \int_{t_0}^{t} |\gamma_j|^2 ds$$

$$- \sum_{k=j+1}^{n-1} \frac{\delta_j \delta_k L_j(t_0) L_k(t_0)}{1 + \delta_k L_k(t_0)} f'(\delta_j L_j(t_0)) \int_{t_0}^{t} \gamma_j^\top \gamma_k ds$$

$$+ f'(\delta_j L_j(t_0))\delta_j L_j(t_0) \int_{t_0}^{t} \gamma_j^\top dW^{(n)}, \quad 1 \leq j < n, \qquad (6.10)$$

and so are deterministic and Gaussian, respectively. The next Picard-Lindelof iteration, however, will be non-Gaussian in general.

By using the approximations $Z_j^{(0)}$ for the expression (6.8) in (6.3), we find a lognormal approximation which we call the (g)-approximation,

$$\ln \frac{L_i^{(g)}(t)}{L_i(t_0)} := - \int_{t_0}^t \frac{|\gamma_i(s)|^2}{2} ds + \int_{t_0}^t \gamma_i^\top(s) dW^n(s)$$

$$- \sum_{j=i+1}^{n-1} \int_{t_0}^t \frac{\delta_j L_j(t_0) \gamma_i^\top \gamma_j(s)}{1 + \delta_j L_j(t_0)} ds \qquad (6.11)$$

which turns out to be a considerable path-wise improvement of the (0)-approximation and is suggested in Brace, Gatarek & Musiela (1997). See also Schoenmakers & Coffey (1998) for several applications, for instance, a multi-factor generalization of the swaption approximation in Brace, Gatarek & Musiela (1997).

By expanding f, f', and f'' as $f(x) = x + \mathcal{O}(x^2)$, $f'(x) = 1 + \mathcal{O}(x)$, and $f''(x) = -2 + \mathcal{O}(x)$, respectively and denoting identity modulo terms of order $\mathcal{O}(\delta_j^2 L_j^2(t_0))$ by $\cong$, we have

$$Z_j^{(1)}(t) \cong \delta_j L_j(t_0) \left(1 + \int_{t_0}^t \gamma_j^\top dW^{(n)} \right)$$

$$- \sum_{k=j+1}^{n-1} \frac{\delta_j \delta_k L_k(t_0) L_j(t_0)}{1 + \delta_k L_k(t_0)} \int_{t_0}^t \gamma_j^\top \gamma_k ds =: \tilde{Z}_j(t). \qquad (6.12)$$

We note that, while in (6.12) the terms in the sum are of order $\mathcal{O}(\delta_j^2 L_j^2(t_0))$, their sum is possibly of order $\mathcal{O}(\delta_j L_j(t_0))$ and therefore not neglected. Now, by substituting $\tilde{Z}$ for (6.8) in (6.3) we get another lognormal approximation, the $(g1)$-approximation,

$$\ln \frac{L_i^{(g1)}(t)}{L_i(t_0)} = - \sum_{j=i+1}^{n-1} \int_{t_0}^t \tilde{Z}_j(s) \gamma_i^\top \gamma_j(s) ds - \frac{1}{2} \int_{t_0}^t |\gamma_i(s)|^2 ds + \int_{t_0}^t \gamma_i^\top(s) dW^{(n)}(s).$$

$$(6.13)$$

The $(g1)$-approximation in its turn improves the (g)-approximation significantly as will appear from a comparative analysis in (6.2.2). We will also consider a simplification of the $(g1)$-approximation, the $(g1')$-approximation, which is obtained by dropping the sum-term in (6.12).

Finally, by substituting $Z^{(1)}$ for (6.8) in (6.3), so without linearization of the function f and its first and second derivative, we get a further refinement, the $(g2)$-approximation, given by

$$\ln \frac{L_i^{(g2)}(t)}{L_i(t_0)} = - \sum_{j=i+1}^{n-1} \int_{t_0}^t Z_j^{(1)}(s) \gamma_i^\top \gamma_j(s) ds - \frac{1}{2} \int_{t_0}^t \gamma_i^2(s) ds + \int_{t_0}^t \gamma_i^\top(s) dW^{(n)}(s),$$

$$(6.14)$$

where $Z^{(1)}$ is given by (6.10).

6.2.2   Simulation Analysis of Different Libor Approximations

It will now be interesting to carry out a comparative numerical analysis of the different lognormal approximations. For typical market volatilities we will compare the lognormal approximations path-wise with the "true" solution, obtained by solving the stochastic differential equation (6.2) by the Euler method using small time steps. In addition, we will carry out a comparison of (non-lognormal) approximations obtained by the Euler scheme using larger time steps. For a uniform tenor structure we will experiment with Euler steps equal to δ, 2δ, or 3δ, etc.

For a correct path-wise comparison, we construct all the lognormal approximations in one common probability space by solving their accompanying stochastic differential equations by the Euler scheme with small time steps, namely, the time steps used for the "true" solution. In the numerical schemes, this is easily achieved by using one and the same Wiener increments for all approximations. In fact, we will solve the SDE's for the log-Libors and their approximations rather than the Libors as in this way we attain a path-wise accuracy of order *one* with the Euler scheme, due to the deterministic diffusion coefficients in the several $\log$ –Libor SDE's. For an extensive study of the numerical solution of stochastic differential equations we refer to the works of Kloeden & Platen (1992), Milstein (1995), and Milstein & Tretyakov (2004). The different SDE's for the $\log$ –Libors are straightforwardly obtained by taking the differential form of (6.11), (6.13), and (6.14), respectively.

In our next experiments we take $n = 61$ and $\delta_i = 0.25$, i.e., a rather large tenor structure of fifteen years, along with the same initial conditions as used for the generation of Figures 6.1 and 6.2: $L_i(0) = 0.061$. We take for European markets realistic scalar volatilities, $|\gamma_i| = 0.15$, and for the correlation structure (6.1) we take $\alpha = 0.9$ and $\beta = 0.04$, yielding a more or less realistic correlation matrix (Table 6.1). For comparison see, for instance, the correlation table in Brace, Gatarek & Musiela (1997). Note that the off-diagonals in Table 6.1 are slightly increasing. The "true" solution is simulated by the Euler method using $\Delta t = \delta/5$.[1] In Tables 6.2– 6.5 it is shown how often the relative error (in per cent) of the corresponding approximation to L_{12}, L_{24}, L_{36}, and L_{48} lays in the relevant percentage intervals (there is no table for L_{60} because, of course, L_{60} is exactly lognormal in the terminal measure). The relative error between $\hat{L}_i$, the numerical solution to the original equation (6.2) and, for instance $\hat{L}_i^{(g)}$, the numerical solution to the SDE belonging to (g)-approximation is defined as

$$\epsilon_i^{(g)} = \max_{1 \le j \le i} \frac{|\hat{L}_i(T_j) - \hat{L}_i^{(g)}(T_j)|}{\hat{L}_i(T_j)}. \tag{6.15}$$

---

[1] For this choice of Δt it appeared by comparison with much smaller Δt that the discretization bias of the "true" solution is negligible w.r.t. the bias of the different approximations.

The relative errors to other approximations are defined analogously.

The numbers in the columns 2 - 8 of Table 6.2 show the number of samples out of 100 for which the event shown in the first column happens. First, we conclude that the path-wise errors produced by the (0)-approximation are generally not tolerable as almost every path has an error in the sense of (6.15) of at least 2-3%. The (g)-approximation proposed by Brace, Gatarek & Musiela (1997) and used for their swaption approximation formula there gives a considerable improvement but is clearly outperformed by the lognormal approximations $(g1')$, $(g1)$, and $(g2)$. For smaller to moderate ϵs, $(g2)$ appears to perform slightly better than $(g1)$ and, in its turn, $(g1)$ somewhat better than $(g1')$, as expected; however, for high ϵ (close to 100%) the reverse conclusion can be made, which indicates that $(g2)$ produces higher outliers than $(g1)$ and so on. Obviously, the (non-lognormal) δ-stepping Euler approximation appears to be the most accurate, but, as we will see in Section 6.3, in many applications it is more expensive than the proposed lognormal approximations since the distributions of the latter may be simulated directly at the desired points in time by methods described in that section. In addition, for the typical model parameters chosen in these experiments, we may conclude that the refined lognormal approximations are roughly comparable with 2δ- or 3δ-stepping Euler while the implementation of the latter is in many situations less efficient (see Section 6.3 for details).

REMARK 6.1   As the bias of different Libor approximations with respect to the terminal measure is caused by the approximation of the sum term in (6.3), we may expect that the larger this term is, the larger the bias will be. In particular, this sum term tends to zero when $t \downarrow t_0$ and when $i \uparrow n - 1$. For instance, to estimate roughly which $L_i(T_i)$ has the largest bias in a specific lognormal approximation, we may approximate the sum term in (6.3) by replacing the L_j by their initial values (as in the (g)-approximation) and then obtain for the experiments in Section 6.2.2,

$$\ln \frac{L_i(T_i)}{L_i(0)} = \ln \frac{L_i(i\delta)}{L_i(0)} = \Sigma^{(i)} - \frac{1}{2}|\gamma|^2 i\delta + \int_0^{i\delta} \gamma_i^\top dW^{(n)}, \quad \text{where}$$

$$\Sigma^{(i)} \approx -|\gamma|^2 \frac{\delta^2 L(0)}{1 + \delta L(0)} \sum_{j=i+1}^{n-1} i \frac{e^{\beta(i-1)^\alpha}}{e^{\beta(j-1)^\alpha}} \tag{6.16}$$

and in the experiments, $L(0) = 0.061$, $\alpha = 0.9$, $\beta = 0.04$, $|\gamma| = 0.15$, $\delta = 0.25$, and $n = 61$. Since in practice α is typically close to 1, we now take $\alpha = 1$ in (6.16) for analytical tractability and thus obtain

$$|\Sigma^{(i)}| \approx |\gamma|^2 \frac{\delta^2 L(0)}{1 + \delta L(0)} \frac{1}{e^\beta - 1} i(1 - e^{-(n-1-i)\beta}).$$

Hence, for $\beta = 0$ (so, in fact, for a one-factor model), we have

$$|\Sigma^{(i_0^*)}| \approx |\gamma|^2 \frac{\delta^2 L(0)}{1 + \delta L(0)} i(n - i - 1),$$

which has a maximum

$$|\Sigma^{(i_0^*)}| \approx |\gamma|^2 \frac{\delta^2 L(0)}{1 + \delta L(0)} \frac{(n-1)^2}{4} \quad \text{for} \quad i = i_0^* \approx \frac{n-1}{2},$$

whereas for $\beta \uparrow \infty$ (full de-correlation) it follows by elementary analysis that the $|\Sigma|-$ maximum, say $|\Sigma^{(i_\beta^*)}|$, goes to zero while $i_\beta^* \to n-2$, non-decreasingly. In a particular situation, however, we can search numerically for i_β^*, and for the experiments in Section 6.2.2 we thus find $i_{0.04}^* \approx 37$, which is more or less consistent with the results listed in Tables 6.2– 6.5.   □

6.3   Direct Simulation of Lognormal Approximations

The results given in Tables 6.2– 6.5 show clearly that the lognormal models (0) and (g) are reasonable approximations and that the models ($g1'$), ($g1$), and ($g2$) are pretty good approximations to the solution of SDE (6.2). We now present direct techniques for lognormal approximations, in particular we illustrate an effective simulation of the (g)-approximation by a lognormal random field in Section 6.3.1.

The motivation for direct simulation methods is clear: In contrast to standard numerical solution of stochastic differential equations there is no need to take small time steps. Indeed, it is possible to construct the solution directly at the desired points in time, for example, at the points of the given tenor structure $0 < T_1 < T_2 < \ldots < T_n$.[2] It will be shown in Section 6.4 that in many typical applications direct simulation methods take much less computing time.

6.3.1   Random Field Simulation of the (g)-approximation

Due to the simple correlation structure of the (g)-approximation it is possible to set up a lognormal random field model by a simulation technique as studied in Sabelfeld (1991) in a more general setting. We construct a lognormal random field whose first two statistical moments are consistent with those

---

[2]From now on, (T_i) is an arbitrary sequence of future time points, so T_i is not necessarily equal to $T_i = i\delta$ as in Section 6.2.2.

TABLE 6.1:   Instantaneous correlations $\rho(\Delta L_i, \Delta L_j)$ between different Libors for $\alpha = 0.9$, $\beta = 0.04$

i,j	4	8	12	16	20	24	28	32	36	40	44	48	52	56	60
4	1	0.88	0.79	0.70	0.63	0.57	0.51	0.46	0.42	0.38	0.34	0.31	0.28	0.26	0.23
8	0.88	1	0.89	0.80	0.71	0.64	0.58	0.52	0.47	0.43	0.39	0.35	0.32	0.29	0.26
12	0.79	0.89	1	0.89	0.80	0.72	0.65	0.59	0.53	0.48	0.43	0.39	0.36	0.32	0.29
16	0.70	0.80	0.89	1	0.90	0.81	0.73	0.66	0.59	0.54	0.49	0.44	0.40	0.36	0.33
20	0.63	0.71	0.80	0.90	1	0.90	0.81	0.73	0.66	0.60	0.54	0.49	0.44	0.40	0.37
24	0.57	0.64	0.72	0.81	0.90	1	0.90	0.81	0.73	0.66	0.60	0.55	0.49	0.45	0.41
28	0.51	0.58	0.65	0.73	0.81	0.90	1	0.90	0.82	0.74	0.67	0.61	0.55	0.50	0.45
32	0.46	0.52	0.59	0.66	0.73	0.81	0.90	1	0.90	0.82	0.74	0.67	0.61	0.55	0.50
36	0.42	0.47	0.53	0.59	0.66	0.73	0.82	0.90	1	0.90	0.82	0.74	0.67	0.61	0.56
40	0.38	0.43	0.48	0.54	0.60	0.66	0.74	0.82	0.90	1	0.91	0.82	0.74	0.68	0.61
44	0.34	0.39	0.43	0.49	0.54	0.60	0.67	0.74	0.82	0.91	1	0.91	0.82	0.75	0.68
48	0.31	0.35	0.39	0.44	0.49	0.55	0.61	0.67	0.74	0.82	0.91	1	0.91	0.82	0.75
52	0.28	0.32	0.36	0.40	0.44	0.49	0.55	0.61	0.67	0.74	0.82	0.91	1	0.91	0.82
56	0.26	0.29	0.32	0.36	0.40	0.45	0.50	0.55	0.61	0.68	0.75	0.82	0.91	1	0.91
60	0.23	0.26	0.29	0.33	0.37	0.41	0.45	0.50	0.56	0.61	0.68	0.75	0.82	0.91	1

TABLE 6.2:   The cumulative distribution of the relative error ϵ_{12}, for 100 paths under different approximations

$100 \cdot \epsilon_{12}$	(0)	(g)	$(g1')$	$(g1)$	$(g2)$	δ st. E.	$2 - \delta$ st. E.
$\leq 0.01\%$	0	0	0	0	19	4	0
$\leq 0.02\%$	0	0	3	10	47	34	5
$\leq 0.03\%$	0	1	10	19	77	73	21
$\leq 0.04\%$	0	3	15	34	89	88	41
$\leq 0.05\%$	0	8	24	45	91	99	57
$\leq 0.06\%$	0	9	32	62	93	100	71
$\leq 0.07\%$	0	12	46	79	93	100	75
$\leq 0.08\%$	0	13	56	99	96	100	82
$\leq 0.09\%$	0	14	71	100	98	100	88
$\leq 0.1\%$	0	17	93	100	100	100	93
$\leq 0.2\%$	0	41	100	100	100	100	100
$\leq 0.3\%$	0	62	100	100	100	100	100
$\leq 0.4\%$	0	77	100	100	100	100	100
$\leq 0.5\%$	0	89	100	100	100	100	100
$\leq 0.6\%$	0	95	100	100	100	100	100
$\leq 0.7\%$	0	97	100	100	100	100	100
$\leq 0.8\%$	0	97	100	100	100	100	100
$\leq 0.9\%$	0	100	100	100	100	100	100
$\leq 1\%$	0	100	100	100	100	100	100
$\leq 2\%$	0	100	100	100	100	100	100
$\leq 3\%$	79	100	100	100	100	100	100
$\leq 4\%$	100	100	100	100	100	100	100

TABLE 6.3: The cumulative distribution of the relative error ϵ_{24}, for 100 paths under different approximations

$100 \cdot \epsilon_{24}$	(0)	(g)	$(g1')$	$(g1)$	$(g2)$	δ st. E.	3δ st. E.
$\leq 0.01\%$	0	0	0	0	3	0	0
$\leq 0.02\%$	0	0	0	0	4	17	0
$\leq 0.03\%$	0	0	2	0	14	50	0
$\leq 0.04\%$	0	0	2	4	20	71	6
$\leq 0.05\%$	0	0	4	9	30	91	10
$\leq 0.06\%$	0	0	10	11	37	96	15
$\leq 0.07\%$	0	2	16	16	47	99	23
$\leq 0.08\%$	0	2	17	22	51	99	32
$\leq 0.09\%$	0	2	19	24	57	99	48
$\leq 0.1\%$	0	3	22	29	66	100	53
$\leq 0.2\%$	0	16	59	93	92	100	92
$\leq 0.3\%$	0	22	100	98	96	100	99
$\leq 0.4\%$	0	34	100	100	99	100	100
$\leq 0.5\%$	0	43	100	100	100	100	100
$\leq 0.6\%$	0	47	100	100	100	100	100
$\leq 0.7\%$	0	60	100	100	100	100	100
$\leq 0.8\%$	0	69	100	100	100	100	100
$\leq 0.9\%$	0	73	100	100	100	100	100
$\leq 1\%$	0	75	100	100	100	100	100
$\leq 2\%$	0	98	100	100	100	100	100
$\leq 3\%$	0	100	100	100	100	100	100
$\leq 4\%$	21	100	100	100	100	100	100
$\leq 5\%$	65	100	100	100	100	100	100
$\leq 6\%$	91	100	100	100	100	100	100
$\leq 7\%$	98	100	100	100	100	100	100
$\leq 8\%$	100	100	100	100	100	100	100

TABLE 6.4:   The cumulative distribution of the relative error ϵ_{36}, for 100 paths under different approximations

$100 \cdot \epsilon_{36}$	(0)	(g)	$(g1')$	$(g1)$	$(g2)$	δ st. E.	4δ st. E.
$\leq 0.01\%$	0	0	0	0	0	0	0
$\leq 0.02\%$	0	0	0	0	1	17	0
$\leq 0.03\%$	0	0	0	0	3	49	1
$\leq 0.04\%$	0	0	1	4	4	77	2
$\leq 0.05\%$	0	0	1	6	8	89	4
$\leq 0.06\%$	0	1	4	9	10	91	6
$\leq 0.07\%$	0	1	5	11	13	94	11
$\leq 0.08\%$	0	1	7	13	18	98	14
$\leq 0.09\%$	0	3	10	16	24	99	24
$\leq 0.1\%$	0	3	12	17	30	99	33
$\leq 0.2\%$	0	10	33	45	75	100	84
$\leq 0.3\%$	0	21	54	87	89	100	94
$\leq 0.4\%$	0	25	94	95	92	100	98
$\leq 0.5\%$	0	34	96	95	93	100	100
$\leq 0.6\%$	0	36	96	96	96	100	100
$\leq 0.7\%$	0	40	97	96	96	100	100
$\leq 0.8\%$	0	49	98	97	96	100	100
$\leq 0.9\%$	0	52	98	98	96	100	100
$\leq 1\%$	0	56	99	98	97	100	100
$\leq 2\%$	0	89	100	100	100	100	100
$\leq 3\%$	1	97	100	100	100	100	100
$\leq 4\%$	12	99	100	100	100	100	100
$\leq 5\%$	42	100	100	100	100	100	100
$\leq 6\%$	75	100	100	100	100	100	100
$\leq 7\%$	89	100	100	100	100	100	100
$\leq 8\%$	93	100	100	100	100	100	100
$\leq 9\%$	97	100	100	100	100	100	100
$\leq 10\%$	99	100	100	100	100	100	100

TABLE 6.5:   The cumulative distribution of the relative error ϵ_{48}, for 100 paths under different approximations

$100 \cdot \epsilon_{48}$	(0)	(g)	$(g1')$	$(g1)$	$(g2)$	δ st. E.	4δ st. E.
$\leq 0.01\%$	0	0	0	0	0	6	0
$\leq 0.02\%$	0	0	0	0	0	55	0
$\leq 0.03\%$	0	0	1	0	0	82	0
$\leq 0.04\%$	0	0	2	1	6	90	4
$\leq 0.05\%$	0	0	2	4	9	95	12
$\leq 0.06\%$	0	0	3	4	13	97	21
$\leq 0.07\%$	0	0	7	8	14	97	31
$\leq 0.08\%$	0	0	10	12	20	98	41
$\leq 0.09\%$	0	0	15	19	21	98	52
$\leq 0.1\%$	0	0	16	22	26	99	67
$\leq 0.2\%$	0	8	40	46	71	100	89
$\leq 0.3\%$	0	17	74	85	83	100	97
$\leq 0.4\%$	0	22	93	93	91	100	99
$\leq 0.5\%$	0	30	95	94	92	100	100
$\leq 0.6\%$	0	35	95	95	95	100	100
$\leq 0.7\%$	0	40	96	95	95	100	100
$\leq 0.8\%$	0	41	96	96	95	100	100
$\leq 0.9\%$	0	45	96	96	96	100	100
$\leq 1\%$	0	54	96	96	96	100	100
$\leq 2\%$	2	90	100	100	100	100	100
$\leq 3\%$	19	97	100	100	100	100	100
$\leq 4\%$	55	98	100	100	100	100	100
$\leq 5\%$	79	100	100	100	100	100	100
$\leq 6\%$	88	100	100	100	100	100	100
$\leq 7\%$	95	100	100	100	100	100	100
$\leq 8\%$	97	100	100	100	100	100	100
$\leq 9\%$	99	100	100	100	100	100	100
$\leq 10\%$	100	100	100	100	100	100	100

of the (g)-approximation. We thus introduce a lognormal random model

$$\mathcal{L}^{(g)}(i,t) = \exp[\xi^{(g)}(i,t)] \tag{6.17}$$

in the measure P_n, where the $\xi^{(g)}(i,t)$, $i = 1,\ldots,n-1$, $t_0 \le t \le T_i$, are Gaussian, such that mean and covariance structure coincide with that of $\ln(L_i^{(g)}(t)/L_i(t_0))$, $t_0 \le t \le T_i$, $i = 1,\ldots n-1$. Hence,

$$E_n\,\xi^{(g)}(i,t) = E_n\,\ln\frac{L_i^{(g)}(t)}{L_i(t_0)}, \tag{6.18}$$

$$E_n\,\xi^{(g)}(i_1,t_1)\xi^{(g)}(i_2,t_2) = E_n\,\Big(\ln\frac{L_{i_1}^{(g)}(t_1)}{L_{i_1}(t_0)}\Big)\Big(\ln\frac{L_{i_2}^{(g)}(t_2)}{L_{i_2}(t_0)}\Big). \tag{6.19}$$

From (6.11) we see that

$$E_n\,\xi^{(g)}(i,t) =: \mu^{(g)}(i,t;t_0) \tag{6.20}$$

$$= -\sum_{j=i+1}^{n-1}\frac{\delta_j L_j(t_0)}{1+\delta_j L_j(t_0)}\int_{t_0}^{t}\gamma_i^\top\gamma_j(s)ds - \frac{1}{2}\int_{t_0}^{t}|\gamma_i|^2(s)ds,$$

where in definition (6.20), t_0 is included as a fixed parameter in the expectation function $\mu^{(g)}$. Further, by (6.11) we have (with $\wedge$ denoting minimum),

$$E_n\,\xi^{(g)}(i_1,t_1)\xi^{(g)}(i_2,t_2) = \mu^{(g)}(i_1,t_1;t_0)\mu^{(g)}(i_2,t_2;t_0)+\mathrm{cov}^{(g)}(i_1,i_2,t_0,t_1\wedge t_2), \tag{6.21}$$

where

$$\mathrm{cov}^{(g)}(i_1,i_2,s,t) := \int_{s}^{t}\gamma_{i_1}^\top\gamma_{i_2}(s)ds, \quad t_0 \le s \le t.$$

We now construct numerically the desired random field $\mathcal{L}^{(g)}(i,T_j)$, $1 \le i < n$, $1 \le j \le i$. To do this, we could simulate the Gaussian vector with the given covariance structure by a conventional simulation technique; see for instance Sabelfeld (1991). However, the specific time correlation (6.21) resulting from the fact that $\xi^{(g)}$ has independent increments suggests a different simulation algorithm.

Indeed, in the first step, we simulate a $n-1$-dimensional Gaussian vector $[\xi^{(g)}(1,T_1),\,\ldots,\,\xi^{(g)}(n-1,T_1)]$ as

$$\xi^{(g)}(i,T_1) = \mu^{(g)}(i,T_1;t_0) + \sum_{k=1}^{K_1}h_{ik}^{(1)}\eta_k^{(1)}, \quad i = 1,\ldots,n-1, \tag{6.22}$$

where the $(n-1) \times K_1$ matrix $[h_{ik}^{(1)}]$ satisfies a Cholesky decomposition

$$\sum_{k=1}^{K_1}h_{ik}^{(1)}h_{jk}^{(1)} = \mathrm{cov}^{(g)}(i,j,t_0,T_1), \quad i,j = 1,\ldots,n-1, \tag{6.23}$$

$h_{ik}^{(1)} = 0$ for $i > n - 1 - K_1 + k$, $\{\eta_k^{(1)}\}_{k=1}^{K_1}$ is a set of independent standard Gaussian random numbers, and K_1 is the rank of the covariance matrix (6.23). Note that, in general, $K_1 \leq n - 1$ and $K_1 = d$, the number of independent Brownian motions in the Libor model, in the case where the γ_i are time independent.

In the l-th step $(2 \leq l \leq n - 1)$ we have

$$\xi^{(g)}(i, T_l) = \xi^{(g)}(i, T_{l-1})\mu^{(g)}(i, T_l; t_0) - \mu^{(g)}(i, T_{l-1}; t_0)$$
$$+ \sum_{k=1}^{K_l} h_{ik}^{(l)}\eta_k^{(l)}, \quad i = l, \ldots, n - 1, \tag{6.24}$$

where the $(n - l) \times K_l$ matrix $[h_{ik}^{(l)}]_{i,k=l,\ldots,n-1}$ satisfies a decomposition

$$\sum_{k=1}^{K_l} h_{ik}^{(l)}h_{jk}^{(l)} = \text{cov}^{(g)}(i, j, T_{l-1}, T_l), \quad i, j = l, \ldots, n - 1, \tag{6.25}$$

$h_{ik}^{(l)} = 0$ for $i > n - 1 - K_l + k$, $\{\eta_k^{(l)}\}_{k=1}^{K_l}$ is a set of independent standard Gaussian random numbers, and K_l is the rank of the covariance matrix (6.25). So, in general, $K_l \leq n - l$ and $K_l \leq \min(d, n - l)$ in the case where the γ_i are time independent.

After $n - 1$ steps we thus find

$$L_i^{(g)}(T_j) = L_i(t_0)\mathcal{L}^{(g)}(i, T_j) \tag{6.26}$$
$$= L_i(t_0)\exp[\xi^{(g)}(i, T_j)], \quad i = 1, \ldots, n - 1, j = 1, \ldots, i.$$

Analogously, the same procedure could be easily carried out for the lognormal approximations (0). Indeed, the simulation formulae (6.22)-(6.26) remain the same, but for the (0)-approximation the functions $\mu^{(g)}$ and $cov^{(g)}$ should be replaced with

$$\mu^{(0)}(i, t; t_0) \equiv 0, \quad \text{and} \quad \text{cov}^{(0)}(i_1, i_2, s, t) = \text{cov}^{(g)}(i_1, i_2, s, t), \quad t_0 \leq s \leq t.$$

REMARK 6.2   Of course, in a random field Monte Carlo simulation all the Cholesky decompositions above can be computed outside of the simulation loop. In general, the cost of the Cholesky decomposition in the l-th step of the random field construction is $\mathcal{O}((n - l)^3)$, so the total cost of the different Cholesky decomposition is $\mathcal{O}(n^4)$. Moreover, in the case where the γ_i are *time independent* it is easily seen that, in fact, only one Cholesky decomposition has to be computed at a cost of $\mathcal{O}(n^3)$. However, for a detailed cost comparison of random field simulation to other direct simulation methods and Euler stepping SDE simulation, see Section 6.4.1. ▯

6.3.2   Simulation of the $(g1')$, $(g1)$, and $(g2)$-approximation

Let us consider the $(g1')$-approximation by example, as the $(g1)$ and $(g2)$ can be treated analogously. For getting the $(g1')$-approximation we have to plug (6.12), with the sum term ignored, into (6.13). Similar to (6.17) we then introduce a lognormal random model

$$\mathcal{L}^{(g1')}(i,t) = \exp[\xi^{(g1')}(i,t)] \tag{6.27}$$

in the measure P_n and define the expectation function

$$E_n\,\xi^{(g1')}(i,t) = E_n\,\ln \frac{L_i^{(g1')}(t)}{L_i(t_0)} =: \mu^{(g1')}(i,t;t_0)$$

$$= -\sum_{j=i+1}^{n-1} \delta_j L_j(t_0) \int_{t_0}^{t} \gamma_i^\top \gamma_j(s)ds - \frac{1}{2}\int_{t_0}^{t} |\gamma_i|^2(s)ds.$$

We now derive from (6.13) by a Fubini argument,

$$\ln \frac{L_i^{(g1')}(t)}{L_i(t_0)} = \mu^{(g1')}(i,t;t_0)$$

$$+ \int_{t_0}^{t} \left(1 - \sum_{j=i+1}^{n-1} \delta_j L_j(t_0) \int_{s}^{t} \gamma_j^\top \gamma_i(u)du\right) \gamma_i^\top(s)dW^{(n)}(s),$$

and thus obtain the covariance function,

$$\mathrm{Cov}(\xi^{(g1')}(i_1,t_1),\xi^{(g1')}(i_2,t_2)) = \int_{t_0}^{t_1 \wedge t_2} ds\,\gamma_{i_1}^\top \gamma_{i_2}(s) \times \tag{6.28}$$

$$\left(1 - \sum_{j=i_1+1}^{n-1} \delta_j L_j(t_0) \int_{s}^{t_1} \gamma_j^\top \gamma_{i_1}(u)du\right)\left(1 - \sum_{k=i_2+1}^{n-1} \delta_k L_k(t_0) \int_{s}^{t_2} \gamma_k^\top \gamma_{i_2}(u)du\right).$$

For the $(g1)$ and $(g2)$-approximation one can derive similar approximations straightforwardly from (6.13), (6.14), respectively.

Unfortunately, the covariance functions of $(g1)$, $(g1')$, and $(g2)$ do not have the special structure that the (0) and (g)-approximation do, so a random field construction as in Section 6.3.1 does not work. For instance, the increment $\ln L_i(T_l) - \ln L_i(T_{l-1})$ is now in general correlated with $\ln L_i(T_{l-1})$ for $i \geq l-1$. However, it is possible to simulate the desired log-Libors simultaneously as one q-dimensional random vector ξ. Let the index set $\mathcal{I}$ be the collection of pairs (i,j) for which $L_i(T_j)$, $1 \leq j \leq i \leq n-1$ is to be simulated. So, q is equal to the number of elements of $\mathcal{I}$ and, for instance, $q = n(n-1)/2$ in case Libors over the whole tenor structure are required. Let further $\phi : \mathcal{I} \longrightarrow \{1,\ldots,q\}$ be an arbitrary bijection, then

$$\xi_{\phi(i,j)} := \ln \frac{L_i^{(\cdot)}(T_j)}{L_i(t_0)} = \mu^{(\cdot)}(i;t_0,T_j) + h_{\phi(i,j),1}\eta_1 + \cdots + h_{\phi(i,j),K}\eta_K, \quad (i,j) \in \mathcal{I},$$

$$\tag{6.29}$$

where $(\cdot)$ stands for $(g1')$, $(g1)$, $(g2)$, respectively, the $q \times K$ matrix h satisfies a Cholesky decomposition

$$\sum_{k=1}^{K} h_{pk} h_{lk} = \text{cov}^{(\cdot)}(\xi_p, \xi_l) \quad p, l = 1, \ldots, q, \tag{6.30}$$

$h_{pk} = 0$ for $p > q - K + k$, $\{\eta_k\}_{k=1}^{K}$ is a set of independent standard Gaussian random numbers, and K is the rank of the covariance matrix (6.30) which, for a specific lognormal approximation, is determined by its covariance structure, for instance, (6.28).

For a full tenor structure, hence for $\mathcal{O}(n^2)$ log-Libors, the Cholesky decomposition will now require a computational cost of $\mathcal{O}(n^6)$ and compared to this the cost of the drift terms can be ignored. It is important to note, however, that all these computations can be done outside of the simulation loop.

6.3.3 Cost Analysis of Euler SDE Simulation and Direct Simulation Methods

Here we give formulae for the cost of Euler SDE simulation, random field simulation of the (g)-approximation, and the direct simulation method for the other lognormal approximations. We disregard all computations which can be done outside of the Monte Carlo simulation loop, such as the computation of various Cholesky decompositions etc.

Let us suppose that we are faced with the simulation of

$$L_i(T_j), \ 1 \le j \le m; \ j \le i \le n - 1, \tag{6.31}$$

for fixed m in the measure P_n.

Euler scheme for solving the log-Libor SDE system

$$dY_i = -\sum_{j=i+1}^{n-1} \frac{\delta_j e^{Y_j}}{1 + \delta_j e^{Y_j}} \gamma_i^\top \gamma_j \, dt - \frac{1}{2} |\gamma_i|^2 dt + \gamma_i^\top dW^{(n)}, \quad i = 1, \ldots, n - 1,$$
$$\tag{6.32}$$

where $Y_i := \ln L_i(t)$. For the volatilities $\gamma_i = (\gamma_{i,1}, \ldots, \gamma_{i,d})$, $1 \le i \le n - 1$, we may assume that $\gamma_{i,k} = 0$ for $i > n - 1 - d + k$, and then it is not difficult to verify that the computation of a single Euler step from t to $t + \Delta t$, $t < T_1$, incurs a cost of

$$Cost_{\text{Euler step}}(n, d, 1) = \tag{6.33}$$

$$\overbrace{(n-2)(c_* + c_{\div} + c_+ + c_{\exp}) + \frac{(n-2)(n-1)}{2}(c_* + c_+)}^{\text{Computation drift term}}$$

$$\overbrace{+d(c_{\text{rand}} + c_*) + [(n-1-d)d + \frac{d^2+d}{2}](c_* + c_+)}^{\text{Computation noise term}}$$

$$=: (n-2)\tilde{c}_{\exp} + [\frac{(n-2)(n-1)}{2} + nd - \frac{d^2+d}{2}]\tilde{c}_* + d\tilde{c}_{\text{rand}},$$

where the cost of one addition, multiplication, division, exponentiation, and the generation of a standard Gaussian random number is denoted by c_+, c_*, $c_{\div}$, $c_{\exp}$, and c_{rand}, respectively. In practice $\tilde{c}_* \approx c_*$, $\tilde{c}_{\exp} \approx c_{\exp}$, $\tilde{c}_{\text{rand}} \approx c_{\text{rand}}$. From (6.33) it is obvious that we have in general for $T_{i-1} < t < T_i$, $i \geq 1$, (where $T_0 := t_0$)

$$Cost_{Eulerstep}(n, d, i) = \tag{6.34}$$

$$(n-i-1)\tilde{c}_{\exp} + \left[\frac{(n-i-1)(n-i)}{2}\right.$$

$$+ \max(n-i-d, 0)d + \frac{\min^2(d, n-i) + \min(d, n-i)}{2}\right]\tilde{c}_*$$

$$+ \min(d, n-i)\tilde{c}_{\text{rand}} =: \mathcal{C}(d, n-i).$$

REMARK 6.3   In our applications n is typically large (e.g., $n \approx 40, 60$) and therefore we could try to deal with asymptotic expression for the behaviour of $\mathcal{C}(d, k)$ for large k and certain d. However, we have to be careful; the cost of the exponential function and the Gaussian random number generator is considerably higher than the cost of a multiplication. For the compiler we used we found by experiment $c_{\exp}/c_* \approx 25$, and with this compiler we found for the random number generator we used, $c_{\text{rand}}/c_* \approx 10$. So in a typical situation, for instance $k \approx 40, 60$, the term involving $c_{\exp}$ in (6.34) can not be ignored at all, and the same is true for the term involving c_{rand} when $d = n - 1$ (a full factor model). Therefore, it is important to consider (6.34) for all k, rather than for $k \to \infty$ only.   ⬚

Obviously, (6.34) yields

$$\mathcal{C}(d, k) = (k-1)\tilde{c}_{\exp} + k^2\tilde{c}_* + k\tilde{c}_{\text{rand}}, \quad k < d,$$

$$\mathcal{C}(d, k) = (k-1)\tilde{c}_{\exp} + [\frac{(k-1)k}{2} + (k-d)d + \frac{d^2+d}{2}]\tilde{c}_* + d\tilde{c}_{\text{rand}}, \quad k \geq d.$$

$$\tag{6.35}$$

The numerical experiments in (6.2.2) have shown that, in practice, for a uniform tenor structure it is accurate enough to take time steps of order $\Delta t = \delta, 2\delta$ for time t up to T_1 and between two tenors T_i, T_{i+1}, we take δ_i for the Euler step size. So, it is clear that the total cost of a thus organized SDE simulation of one sample of the values (6.31) will be equal to

$$
Cost_{\text{SDE}}(n, d, m) = \frac{T_1}{\Delta t} Cost_{\text{Euler step}}(n, d, 1) +
$$

$$
\sum_{i=2}^{m} Cost_{\text{Euler step}}(n, d, i) + Cost_{\text{Exp. calls}}^{(m)}, \quad (6.36)
$$

where $Cost_{\text{Exp. calls}}^{(m)}$ takes into account the cost of the exponential calls at T_m to obtain Libors rather than log-Libors. For $i < m$ these exponential calls are already included in the first and second term of (6.36) as Libors at T_i are needed in the drift terms. Hence,

$$
Cost_{\text{SDE}}(n, d, m) = \frac{T_1}{\Delta t} C(d, n-1) + \sum_{k=n-m}^{n-2} C(d, k) + (n-m)\tilde{c}_{\exp} \quad (6.37)
$$

with empty sums defined as zero. We derive from (6.35) and (6.37) by elementary algebra[3] explicit expressions for the cost of the Euler method for $m > n - d$ and $m \leq n - d$, respectively,

$$
Cost_{\text{SDE}}(n, d, m) = \quad (6.38)
$$

$$
\frac{T_1}{\Delta t} \left\{ (n-2)\tilde{c}_{\exp} + \left[\frac{(n-2)(n-1)}{2} + nd - \frac{d^2 + d}{2} \right] \tilde{c}_* + d\tilde{c}_{\text{rand}} \right\}
$$

$$
+ \left\{ (n-1)m - \frac{m^2 + 3m - 4}{2} \right\} \tilde{c}_{\exp}
$$

$$
+ \left\{ -\frac{n^3}{6} + \frac{2m + d - 1}{2} n^2 - \frac{6m^2 + 6m + 3d^2 + 6d - 10}{6} n \right.
$$

$$
\left. + \frac{2m^3 + 3m^2 + m + d^3 + 3d^2 + 2d - 6}{6} \right\} \tilde{c}_*
$$

$$
+ \left\{ -\frac{n^2}{2} + \frac{2m + 2d + 1}{2} n - \frac{m^2 + m + d^2 + 3d}{2} \right\} \tilde{c}_{\text{rand}}, \quad m > n - d
$$

---

[3] The expressions (6.38) and (6.39) can be checked easily with Mathematica or Maple, for instance.

and

$$Cost_{\text{SDE}}(n, d, m) =$$

$$\frac{T_1}{\Delta t}\left\{(n-2)\tilde{c}_{\exp} + \left[\frac{(n-2)(n-1)}{2} + nd - \frac{d^2+d}{2}\right]\tilde{c}_* + d\tilde{c}_{\text{rand}}\right\}$$

$$+\left\{(n-1)m - \frac{m^2+3m-4}{2}\right\}\tilde{c}_{\exp} + \left\{\frac{m-1}{2}n^2 - \frac{m^2+2m-2md+2d-3}{2}n\right.$$

$$\left.+\frac{m^3+3m^2+2m-6-3d^2m-3dm^2+3d^2+3d}{6}\right\}\tilde{c}_*$$

$$+(m-1)d\tilde{c}_{\text{rand}}, \quad m \leq n - d. \tag{6.39}$$

The expressions (6.38) and (6.39) will be used later for a cost comparison of random field simulation and direct simulation of different lognormal approximations with Euler SDE simulation in various situations.

Random field simulation technique

Here we consider the general case where γ is time dependent and we thus have to take in (6.24), $K_l = n - l$ for $1 \leq l \leq n$, even if d is small. We choose the $(n-l) \times (n-l)$-matrix $h^{(l)}$ in (6.24) as an upper-triangular matrix. Disregarding again pre-computation costs outside of the Monte Carlo loop, it follows from (6.24) that the cost of calculating one sample of the Libors (6.31) is given by

$$Cost_{\text{RFS}}(n, m) = \sum_{l=1}^{m}\left\{\frac{1}{2}(n-l)(n-l+1)\tilde{c}_* + (n-l)\tilde{c}_{\text{rand}}\right\} + Cost_{\text{Exp. calls}}^{(m,n)}$$

$$= \frac{1}{6}(3mn^2 - 3m^2n + m^3 - m)\tilde{c}_* + \frac{1}{2}m(2n - m - 1)(\tilde{c}_{\text{rand}} + \tilde{c}_{\exp}),$$

where the term $Cost_{\text{Exp. calls}}^{(m,n)} = \frac{1}{2}m(2n - m - 1)\tilde{c}_{\exp}$ is taken into account because we need Libors rather than log-Libors.

REMARK 6.4   Let us consider a volatility structure which is piecewise constant on the interval $[T_1, T_n]$, hence $\gamma_i(s) = \gamma_i^{(p)}$ for $T_p \leq s < T_{p+1}$, $1 \leq p < n$. In this situation the first Cholesky decomposition in the random field construction (6.23) might have rank $n-1$, but for the Cholesky decompositions in step $2 \leq l < n$ we may take $K_l = \min(d, n - l)$. As a consequence, the random field simulation will be much faster in case d is small. Indeed, it is easily verified that for the computation cost we now have

$$Cost_{\text{RFS}}^{\gamma\,p.c.[T_1,T_n]}(n,d,m) \leq \frac{1}{2}n(n-1)\tilde{c}_* + (n-1)\tilde{c}_{\text{rand}} +$$

$$\sum_{l=2}^{m}\left[(n-l)d\tilde{c}_* + d\tilde{c}_{\text{rand}}\right] + Cost_{\text{Exp. calls}}^{(m,n)}$$

$$= \frac{1}{2}n(n-1)\tilde{c}_* + (n(m-1) - \frac{m^2+m-2}{2})d\tilde{c}_*$$

$$+ (n+md-d-1)\tilde{c}_{\text{rand}} + \frac{1}{2}m(2n-m-1)\tilde{c}_{\text{exp}}.$$

$$(6.40)$$

Even for a full tenor structure $m = n-1$, we have

$$Cost_{\text{RFS}}^{\gamma\,p.c.[T_1,T_n]}(n,d,m=n-1) \leq (\frac{n^2-n}{2} + \frac{n^2-3n+2}{2}d)\tilde{c}_* \qquad (6.41)$$

$$+ (n(d+1) - 2d-1)\tilde{c}_{\text{rand}} + \frac{1}{2}n(n-1)\tilde{c}_{\text{exp}},$$

and so the coefficients of $\tilde{c}_*$, $\tilde{c}_{\text{rand}}$, and $\tilde{c}_{\text{exp}}$ in (6.40) and (6.41) are for small d quadratic and linear in n, respectively, whereas the corresponding coefficients in the cost of the SDE simulation tend to be larger: For example, when $m = n-1$ in (6.38) the coefficient of $\tilde{c}_*$ still contains $n^3/6$ and $\frac{T_1}{\Delta t}\frac{n^2}{2}$ for small d.

If, moreover, γ_i is constant on $[t_0, T_1[$, i.e., $\gamma_i(s) = \gamma_i^{(0)}$ for $t_0 \leq t < T_1$, a rank-d Cholesky decomposition applies to the $\xi^{(g)}(T_1)$ construction also and then we have

$$Cost_{\text{RFS}}^{\gamma\,p.c.[t_0,T_n]}(n,d,m) \leq \sum_{l=1}^{m}[(n-l)d\tilde{c}_* + d\tilde{c}_{\text{rand}}] + Cost_{\text{Exp. calls}}^{(m,n)}$$

$$= (nm - \frac{m^2+m}{2})d\tilde{c}_* + md\tilde{c}_{\text{rand}}$$

$$+ \frac{1}{2}m(2n-m-1)\tilde{c}_{\text{exp}}, \qquad (6.42)$$

which means a next speed up with respect to Euler SDE simulation in the case where $d \ll n$; for example, compare (6.42) with (6.39) for $d = 1$. □

Direct simulation of the $(g1')$, $(g1)$, and $(g2)$-approximation

By taking $q = \sum_{j=1}^{m}(n-j) = \frac{1}{2}m(2n-m-1)$ in (6.29) we see that with an upper-triangular h, for the full rank case $K = q$, the simulation of one sample of (6.31) will cost inside of the loop:

$$Cost_{DS}(n, m) = \frac{1}{2}q(q + 1)\tilde{c}_* + q\tilde{c}_{rand} + Cost_{Exp.\ calls}^{(m,n)}$$

$$= \left\{\frac{m^2}{2}n^2 - \frac{m^3 + m^2 - m}{2}n + \frac{m(m + 1)(m^2 + m - 2)}{8}\right\}\tilde{c}_* +$$

$$+ \frac{1}{2}m(2n - m - 1)(\tilde{c}_{rand} + \tilde{c}_{exp}). \tag{6.43}$$

It is clear that for $m = n - 1$ the first term will be of order $\mathcal{O}(n^4 c_*)$, and so direct simulation of the $(g1')$, $(g1)$, or $(g2)$-approximation for a full tenor structure $(m = n - 1)$ can only be recommended when n is not too large, whereas for larger n this simulation method is recommended for relatively small values of m. For example, $m = 1$ in (6.43) yields

$$Cost_{DS}(n, m = 1) = \frac{n^2 - n}{2}\tilde{c}_* + (n - 1)(\tilde{c}_{rand} + \tilde{c}_{exp}). \tag{6.44}$$

6.4   Efficiency Gain with Respect to SDE Simulation; an Optimal Simulation Program

6.4.1   Simulation Alternatives

Let us suppose that we want to do a Monte Carlo evaluation of a Libor derivative involving Libors specified in (6.31). Rather than full Euler-stepping from the starting date t_0, it may be profitable to simulate $L(T_1)$ using one of the lognormal approximations and then, for instance, proceed with Euler stepping through the remaining tenors, provided, of course, that $L(T_1)$ is well approximated. Alternatively, as long as the (g)-approximation is tolerable, one may apply the random field simulation technique. It is to be expected that for longer dated products (i.e., larger T_1) and in particular when, additionally, m is small (e.g., $m = 1$ in the case of a European style derivative), both alternatives may yield a substantial efficiency gain. We will compute the order of this gain for a "full" tenor structure, i.e., $m = n - 1$, and the European case $m = 1$, where we distinguish between the multi-factor case $d = n - 1$ and the one factor case $d = 1$.

Alternative 1: Lognormal approximation of $L(T_1)$ followed by Euler stepping
As a first alternative to full Euler stepping we consider a simulation algorithm that first simulates with a proper lognormal approximation the Libors $L(T_1)$ and then proceeds with Euler δ-stepping. Using the results of Section 6.3.3 we may straightforwardly compute for this method the efficiency ratio

$$R_{Eff}^{(1)}(n, d, m) := \frac{Cost_{SDE_{[t_0, T_m]}}}{Cost_{DS_{[t_0, T_1]}\&SDE_{[T_2, T_m]}}}$$

and obtain the following results.

Full structure, multi-factor

$$R_{\text{Eff}}^{(1)}(n, n-1, n-1) \approx \frac{(n\frac{T_1}{\Delta t} + \frac{n^2}{2})\tilde{c}_{\exp} + (n\frac{T_1}{\Delta t} + \frac{n^2}{2})\tilde{c}_{\text{rand}} + (n^2\frac{T_1}{\Delta t} + \frac{n^3}{3})\tilde{c}_*}{\frac{n^2}{2}\tilde{c}_{\exp} + \frac{n^2}{2}\tilde{c}_{\text{rand}} + \frac{n^3}{3}\tilde{c}_*}$$

Full structure, one factor

$$R_{\text{Eff}}^{(1)}(n, 1, n-1) \approx \frac{(n\frac{T_1}{\Delta t} + \frac{n^2}{2})\tilde{c}_{\exp} + (\frac{T_1}{\Delta t} + n)\tilde{c}_{\text{rand}} + (\frac{n^2}{2}\frac{T_1}{\Delta t} + \frac{n^3}{6})\tilde{c}_*}{\frac{n^2}{2}\tilde{c}_{\exp} + 2n\tilde{c}_{\text{rand}} + \frac{n^3}{6}\tilde{c}_*}$$

European, multi-factor

$$R_{\text{Eff}}^{(1)}(n, n-1, 1) \approx \frac{(n\frac{T_1}{\Delta t} + n)\tilde{c}_{\exp} + n\frac{T_1}{\Delta t}\tilde{c}_{\text{rand}} + n^2\frac{T_1}{\Delta t}\tilde{c}_*}{n\tilde{c}_{\exp} + n\tilde{c}_{\text{rand}} + \frac{n^2}{2}\tilde{c}_*}$$

European, one factor

$$R_{\text{Eff}}^{(1)}(n, 1, 1) \approx \frac{(n\frac{T_1}{\Delta t} + n)\tilde{c}_{\exp} + \frac{T_1}{\Delta t}\tilde{c}_{\text{rand}} + \frac{n^2}{2}\frac{T_1}{\Delta t}\tilde{c}_*}{n\tilde{c}_{\exp} + n\tilde{c}_{\text{rand}} + \frac{n^2}{2}\tilde{c}_*}$$

Hence, when $\frac{T_1}{\Delta t} \gtrsim n$ it is clear that this alternative will be substantially faster than full Euler stepping.

Alternative 2: Random field simulation
When volatilities are not too high or T_m is not too large the (g)-approximation might be acceptable and then, particularly for a full tenor structure ($m = n - 1$), it might be profitable to use the random field simulation technique. The efficiency ratio

$$R_{\text{Eff}}^{(2)}(n, d, m) := \frac{Cost_{\text{SDE}_{[t_0, T_m]}}}{Cost_{\text{RFS}_{[T_1, T_m]}}}$$

may again be computed using Section 6.3.3 and for $m = n - 1$ we have

Full structure, multi-factor

$$R_{\text{Eff}}^{(2)}(n, n-1, n-1) \approx \frac{(n\frac{T_1}{\Delta t} + \frac{n^2}{2})\tilde{c}_{\exp} + (n\frac{T_1}{\Delta t} + \frac{n^2}{2})\tilde{c}_{\text{rand}} + (n^2\frac{T_1}{\Delta t} + \frac{n^3}{3})\tilde{c}_*}{\frac{n^2}{2}\tilde{c}_{\exp} + \frac{n^2}{2}\tilde{c}_{\text{rand}} + \frac{n^3}{6}\tilde{c}_*}$$

Full structure, one factor

$$R_{\text{Eff}}^{(2)}(n, 1, n-1) \approx \frac{(n\frac{T_1}{\Delta t} + \frac{n^2}{2})\tilde{c}_{\exp} + (\frac{T_1}{\Delta t} + n)\tilde{c}_{\text{rand}} + (\frac{n^2}{2}\frac{T_1}{\Delta t} + \frac{n^3}{6})\tilde{c}_*}{\frac{n^2}{2}\tilde{c}_{\exp} + \frac{n^2}{2}\tilde{c}_{\text{rand}} + \frac{n^3}{6}\tilde{c}_*},$$

from which we conclude that, in general, the efficiency gain with respect to the first simulation alternative is limited: If, in the full rank case, $n^3\tilde{c}_*$ is much larger than $n^2\tilde{c}_{\exp}$, random field simulation is about two times faster as simulation alternative 1. However, for a low factor model with $\gamma(s)$ piece wise constant between T_1 and T_n, it follows from Remark 6.4 that random field

TABLE 6.6:   Values of $R_{\text{Eff}}^{(1)}$ for Example 6.1
($Cost_{\text{SDE}} \times 10^4 \tilde{c}_*$)

Δt	$d=1$	$d=2$	$d=40$
δ	3.66 (11.6)	3.64 (12.0)	3.83 (19.0)
2δ	2.30 (7.0)	2.29 (7.5)	2.39 (11.8)
3δ	1.84 (5.8)	1.84 (6.0)	1.90 (9.4)

simulation is really faster than lognormal approximation followed by Euler-stepping, and also faster than alternative 1. In the one factor case we then have

Full structure, one factor

$$R_{\text{Eff}}^{(2),\ \gamma\ \text{p.c.}}(n,1,n-1) \approx \frac{(n\frac{T_1}{\Delta t}+\frac{n^2}{2})\tilde{c}_{\exp}+(\frac{T_1}{\Delta t}+n)\tilde{c}_{\text{rand}}+(\frac{n^2}{2}\frac{T_1}{\Delta t}+\frac{n^3}{6})\tilde{c}_*}{\frac{n^2}{2}\tilde{c}_{\exp}+2n\tilde{c}_{\text{rand}}+n^2\tilde{c}_*}.$$

For any problem in practice for which alternatives 1 or 2 applies it will be easy to compute or estimate from the expressions in Section 6.3.3 the efficiency gain with respect to Euler-stepping SDE simulation. However, instead of listing many possibilities, we present below two typical situations in which the proposed simulation alternatives improve on standard Euler-stepping.

Example 6.1

Suppose we have to price a relatively long term Libor option with $T_1 = 12$ years, $T_n = 22$ years, three-month periods, and where Libors over the full tenor structure are required for the valuation of the contract ($m = n - 1$). A product of this type is, for instance, the trigger swap; see Section 6.5. So, $\delta = 0.25$, $n = 41$, $m = 40$. For our C++ compiler the relative cost of a standard Gaussian random number and the exponential function is approximately $\tilde{c}_{\text{rand}}/\tilde{c}_*$, ≈ 10 and $\tilde{c}_{\exp}/\tilde{c}_* \approx 25$, respectively.

The relative cost of full Euler δ-stepping from t_0 to T_n with respect to simulation alternative 1 ($(g2)$-approximation of $L(T_1)$ and Euler δ-stepping from T_2 to T_n),

$$R_{\text{Eff}}^{(1)} := \frac{Cost_{\text{SDE}_{[t_0,T_n]}}}{Cost_{\text{DS}_{[t_0,T_1]}\&\text{SDE}_{[T_2,T_n]}}},$$

is computed from expressions in Section 6.3.3 and listed in Table 6.6 for a one- and two-factor model, a multi-factor model, and different Euler step sizes, $\Delta t = \delta, 2\delta, 3\delta$ on $[t_0, T_1]$. The absolute computation costs of the SDE method in terms of 10^4 multiplications are given in brackets.[4]

---

[4]The approximate expressions in this section give nearly the same results.

TABLE 6.7:   Values of $R_{\text{Eff}}^{(1)}$ for Example 6.2
$(Cost_{\text{SDE}} \times 10^4 \tilde{c}_*)$

Δt	$d = 1$	$d = 2$	$d = 40$
δ	39.5 (8.7)	40.6 (8.9)	64.8 (14.3)
2δ	19.9 (4.3)	20.6 (4.5)	32.6 (7.1)
3δ	13.4 (2.9)	13.8 (3.0)	21.9 (4.8)

Example 6.2

We consider the same problem as in Example 6.1 except that now only $L_i(T_1)$, $i = 1, \ldots, n-1$, are required for pricing the product; so now $m = 1$, hence the European case. For instance, the swaption and the callable reverse floater are products of this type; see Section 6.5 for details. The computed cost ratios of direct simulation of the $(g1), (g1')$, or $(g2)$-approximation with respect to Euler-stepping SDE simulation are listed in Table 6.7 for different step sizes Δt and different d. It is clear that the efficiency gain in this case is tremendous. Note that in general when $m = 1$, the simulation costs inside the loop of the (g)-approximation and the $(g1), (g1')$, or $(g2)$-approximation coincide, and so a refined lognormal approximation as a $(g1'), (g1)$, or $(g2)$-approximation should be preferred in any case.   □

6.4.2   An Optimal Simulation Program

Based on the experimental results with respect to the accuracy of different Libor approximations in Section (6.2.2) and the cost comparison between standard Euler δ-stepping and different alternative simulation methods in section (6.4.1), we now propose the following procedure for the price simulation of a derivative structure involving a system (6.31) of Libors, in a given (calibrated) Libor model.

> *Step I:* Compare the accuracy of different lognormal approximation for the given Libor model by using a path simulation program as designed in Section 6.2.2.

> *Step (II):* Decide which approximations are acceptable in view of the accuracy required for the derivative structure and then choose one of the following options:

>> *Case 1:* If the $(g1), (g1')$, or $(g2)$-approximation is tolerable up to $L(T_1)$ and the (g)-approximation is not suitable in the sense that Case 2 below does not apply, generate $L(T_1)$ by a direct simulation method (Section 6.3.2) and proceed to T_m with Euler δ-stepping.

>> *Case 2:* If the (g)-approximation is tolerable over all tenors, the number of factors, d, is small, and $\gamma(s)$ is piecewise constant be-

tween T_1 and T_m, then apply the random field simulation technique
(Section 6.3.1).

Case 3: If Case 1 and 2 do not apply, SDE simulation by δ-stepping,
or if necessary by smaller time steps, will be acceptable in almost
all practical situations.

6.5   Practical Simulation Examples

We now present some test results on the valuation of the swaption and the
trigger swap and discuss the pricing of a callable reverse floater.

6.5.1   European Swaption

Let us consider a standard European payer swaption with maturity T_1,
strike κ, and principal \$1, which gives the right to enter at T_1 into a swap
contract with strike κ. Hence, by Proposition 1.2, the swaption value is given
in the terminal measure P_n by

$$Swpn(t) := Swpn_{1,n;\kappa}(t) = B_n(t)E_n^{\mathcal{F}_t}\frac{(Swap_{1,n;\kappa}(T_1))^+}{B_n(T_1)}, \qquad (6.45)$$

where $Swap_{1,n;\kappa}(T_1)$ denotes the value of the swap contract at T_1. From
(6.45) and the expression for the price of a standard swap contract (1.23), it
follows by a little algebra that the swaption value can be represented as

$$Swpn(t) = B_n(t)E_n^{\mathcal{F}_t}\left[\sum_{j=1}^{n-1}\frac{B_{j+1}(T_1)}{B_n(T_1)}(L_j(T_1) - \kappa)\delta_j\right]^+. \qquad (6.46)$$

In (6.46), the bond ratios $B_{j+1}(T_1)/B_n(T_1)$ can be expressed in Libors by

$$\frac{B_{j+1}(T_1)}{B_n(T_1)} = \prod_{i=j+1}^{n-1}(1 + \delta_i L_i(T_1)). \qquad (6.47)$$

The expressions (6.46) and (6.47) can thus be used for simulation of swaption
prices based on a large enough Monte Carlo sample of the Libor process. The
initial bond value $B_n(t_0)$ has to be derived from the initial present value curve
given at t_0.

In the spirit of Example 6.2 we simulate the value of an at the money
swaption using different Libor approximations and the results are listed in
Table 6.8. To estimate the systematic error in the swaption value due to
a particular approximation, the swaption value is simulated by the log-SDE

TABLE 6.8: Swaption values for different Libor approximations simulated under the same Wiener processes in base points (i.e., $\times 10^{-4}$)

"true value"	δ-step E.	3δ-step E.	$(g2)$	$(g1)$	$(g1')$	(g)	(0)
379.07	379.54	380.71	381.69	379.33	377.28	408.18	516.4
Sys. err. (%)	0.12	0.43	0.69	0.07	-0.47	7.7	36.2

$T_1 = 12$, $n = 41$, $\delta_i \equiv 0.25$, $L_i(0) \equiv 0.06045$, $|\gamma_i| \equiv 0.15$, $\alpha = 0.9$, $\beta = 0.04$, $\kappa = 0.06045$; Number of trajectories 50000, Monte Carlo error (1 standard deviation) ≈ 5 bp. $\approx 1.3\%$, SDEs simulated with $\Delta t = 0.05$

for this approximation, where the same Wiener processes are used for all approximations.

We note that although the $(g2)$-approximation according to Tables 6.2–6.5 seems to be the best path-wise lognormal approximation overall, it produces larger outliers than $(g1)$ and $(g1')$. For this reason, apparently, the $(g2)$-approximation does not give an essentially better swaption approximation than the $(g1)$ gives. Further we note that by using a much faster direct simulation method the "real time" Monte Carlo error can be considerably reduced by increasing the number of simulations. For instance, a run of 500000 direct simulations of the $(g1)$-approximation takes, for this example, about 5 seconds on a 1Ghz computer and gives a Monte Carlo error of about 0.4%, whereas from Table 6.8 (full rank column) we see that even a 3δ-stepping Euler method, which is comparable to $(g1')$ in accuracy, would have taken about 40 times longer.

6.5.2 Trigger Swap

Next we consider a knock-in version of a payer trigger swap with zero spread coupons (see Section 4.1.1 for the general trigger swap). For given trigger levels $K_1, \ldots, K_n$ and strike κ, the contract is specified as follows. As soon as the discretely monitored spot Libor $L_i(T_i)$ crosses a trigger level K_i from below, the contract holder enters obligatory into a payer swap with fixed coupon κ for the remaining period $[T_i, T_n]$, settled at the tenor dates $T_{i+1}, \ldots, T_n$. According to Proposition 1.2 and (1.23), the value of this trigger swap expressed in the P_n-measure is given by

$$Trswp(t) = B_n(t) E_n^{\mathcal{F}_t} \left[\frac{1}{B_n(T_\tau)} \left(1 - B_n(T_\tau) - \kappa \sum_{j=\tau}^{n-1} B_{j+1}(T_\tau)\delta_j \right) \right],$$

$$(6.48)$$

where τ, the trigger index, is given by $\tau := \min_{1 \leq p < n} \{ p \mid L_p(T_p) > K_p \}$. For Monte Carlo simulation of the trigger swap we express the random variable

TABLE 6.9: Trigger swap values for different Libor approximations simulated under the same Wiener processes in base points (i.e., $\times 10^{-4}$)

"true value"	δ-step E.	$(g2)$	$(g1)$	$(g1')$	(g)	(0)
428.51	429.09	431.84	429.46	427.58	461.42	567.14
Sys. err. (%)	0.13	0.77	0.22	-0.22	7.6	32.3

$T_1 = 12$, $n = 41$, $\delta_i \equiv 0.25$, $L_i(0) \equiv 0.06045$, $|\gamma_i| \equiv 0.15$, $\alpha = 0.9$, $\beta = 0.04$, $\kappa = 0.06045$, $K_i \equiv 0.08$; Number of trajectories 25000, Monte Carlo error (1 standard deviation) ≈ 7 bp. $\approx 1.6\%$, SDEs simulated with $\Delta t = 0.05$

inside the expectation of (6.48) in Libors only to obtain the representation

$$Trswp(t) = B_n(t)E_n^{\mathcal{F}_t}\left[-1 + \prod_{i=\tau}^{n-1}(1+\delta_i L_i(T_\tau)) - \kappa \sum_{j=\tau}^{n-1} \delta_j \prod_{i=j+1}^{n-1}(1+\delta_i L_i(T_\tau))\right].$$
(6.49)

As an application of Example 6.1, we simulate the trigger swap value under different Libor approximations, just as in the swaption example (Section 6.5.1). The trigger levels are taken to be equal, $K_i \equiv 0.08$, and the results are given in Table 6.9. Again we see that the $(g2)$-approximation does not give an essentially better result than the $(g1)$- or $(g1')$-approximation for the same reason as in the swaption example. Regarding the rather large systematic error of the (g)-approximation over a 10-year time period for the example trigger swap above, we propose a direct simulation method as in Case 1, Section 6.4.2, by using the $(g1)$-approximation. On our 1GHz computer a run of 100000 simulations of the (full factor) Libor model using this method takes about two minutes and yields a tolerable Monte Carlo error of about 0.8%, whereas a δ-step Euler method takes about 4 times longer according to Table 6.6. Moreover, although the systematic error of the δ-step Euler method is smaller than the error of the $(g1)$-approximation, the latter error is still much smaller than the Monte Carlo error in this case and so the δ-step Euler method will not give an essentially better result.

6.5.3 Callable Reverse Floater

In particular, for products where Example 6.2 applies the efficiency gain of the proposed simulation method compared with Euler stepping is quite big. A realistic example of such a product is a T_1-callable reverse floater, which is a European call option on a reverse floater at T_1. Generally, reverse floaters occur in different variations; see Section 4.1.5 for further details and properties. Let us take, as a special example, $\kappa' = 0$ in the definition in Section 4.1.5. Then the cash-flows of the reverse floater are given by

$$C_{T_{i+1}} := \delta_i L_i(T_i) - \delta_i \max(\kappa - L_i(T_i), 0), \quad 1 \leq i \leq n-1.$$

According to Section 4.1.5, the price of the reverse floater at time $t \leq T_1$ is given by

$$RF(t) = B_1(t) - B_n(t) - \sum_{i=1}^{n-1} B_{i+1}(t) E_{i+1}^{\mathcal{F}_t} [\delta_i (\kappa - L_i(T_i))^+], \quad (6.50)$$

and can be evaluated analytically in a Libor market model, where $L_i(T_i)$ is log-normally distributed under P_{i+1}. As a consequence, $RF(T_1)/B_n(t)$ may be expressed explicitly as a function of $L(T_1)$, and so it is possible to compute the price of the T_1-callable reverse floater,

$$CRF(t) := B_n(t) E_n^{\mathcal{F}_t} \left[\frac{RF(T_1)^+}{B_n(T_1)} \right], \quad (6.51)$$

by Monte Carlo simulation of $L(T_1)$ in the terminal measure. Hence Example 6.2 applies. Moreover, the efficiency gain for this product will still be large for medium-term maturities T_1. Of course, these conclusions apply also for the case $\kappa' \neq 0$.

6.6 Summarisation and Final Remarks

From Tables 6.2–6.5 in Section 6.2.2, and the Examples 6.5.1 and 6.5.2, we conclude that application of the lognormal (g)-approximation as proposed by Brace, Gatarek and Musiela (1997) in the valuation of long dated derivative structures may lead to intolerable errors in the option values. In Section 6.2 different lognormal approximation were constructed; in particular, besides the (0) and (g)-approximation, we considered the $(g1')$, $(g1)$, and $(g2)$-approximations. Although the (g)-approximation is not reliable for longer-maturity structures, the refined lognormal approximations $(g1')$, $(g1)$, and $(g2)$ are in most cases still acceptable in the sense that the systematic errors in the option values caused by these approximations are well within a Monte Carlo error of about 1%, which in practice is thoroughly tolerable for over the counter options. In view of the larger outliers produced by the $(g2)$-approximation, however, the $(g1)$-approximation appears to be the best candidate in practice, and its implementation according to one of the simulation strategies outlined in Section 6.4.2 turns out to be very efficient compared to standard simulation of the log-Libor SDEs with time steps δ, 2δ, or 3δ. Moreover, the efficiency gain becomes very considerable when this simulation method is applied to products which fit into Example 6.2, in particular European-style derivatives such as the callable reverse floater. For European Libor products, the lognormal approximations can also be used for Monte Carlo evaluation of the concerning hedge parameters (see for instance Milstein & Schoenmakers (2002)). For practical applications we suggest the implementation of a path comparison method as described in Section 6.2.2 as a measuring instrument that helps to decide in a particular situation whether a certain lognormal approximation is acceptable or not. Finally, it should be noted that in this chapter smile effects are not taken into account and the extension of the methods presented here to extended Libor market models (see for instance Section 1.3.4, Andersen & Andreasen (1998)), in order to incorporate volatility skews, would be an interesting subject of future research.

Appendix A

Appendix

A.1 Glossary of Stochastic Calculus

For a detailed exposition of basic concepts and results on probability theory and stochastic processes we refer to standard text books, for example, Kallenberg (2002), Karatzas & Shreve (1991), Ikeda & Watanabe (1981), Revuz & Yor (1998). In this appendix we give a concise collection of necessary stochastic tools used in this book.

A.1.1 Stochastic Processes in Continuous Time

In this appendix, $(\Omega, \mathcal{F}, P)$ is a probability space with filtration $\mathbb{F} = (\mathcal{F}_t)_{0 \leq t \leq T_\infty \leq \infty}$ satisfying the "usual conditions", i.e., continuous from the right, $\bigcap_{s > t} \mathcal{F}_s = \mathcal{F}_t$, and $\mathcal{F}_0$ contains all sets $N \in \mathcal{F}$ with $P(N) = 0$.

A (continuous time) stochastic process X in $\mathbb{R}^m$ is a map $X : \mathbb{R}_+ \times \Omega \ni (t, \omega) \to \mathbb{R}^m$, such that for each $t \geq 0$ the random variable $X(t) : \omega \to X(t, \omega)$ is $\mathcal{F}_t / \mathcal{B}(\mathbb{R}^m)$ measurable, where $\mathcal{B}(\mathbb{R}^m)$ denotes the Borel σ-algebra generated by the open subsets of $\mathbb{R}^m$. If, in addition, the process X is left continuous, it may be called *predictable*.

The mapping

$$(k; t_1, \ldots, t_k; A_1, \ldots A_k) \longrightarrow P\left(\{X(t_1) \in A_1, \ldots, X(t_k) \in A_k\}\right), \quad (A.1)$$

for $k \in \mathbb{N}$, $0 \leq t_1 < \cdots < t_k$, and $A_1, \ldots, A_k$ in $\mathcal{B}(\mathbb{R}^m)$, is called the finite dimensional distribution of the process X. As a general result, the law or distribution of X, being the image probability measure of P on the space of sample paths, is completely determined by the finite dimensional distribution (A.1). Furthermore, the distribution of X is already determined by taking the Borel sets A_i, $i = 1, \ldots, k$, in (A.1) of the form $A_i = \{x \in \mathbb{R}^m : x_p \leq a_p^{(i)}, \quad p = 1, \ldots, m\}$, $a^{(i)} \in \mathbb{R}^m$.

A.1.1.1 Wiener process or Brownian motion

Due to Wiener's theorem there exists an $\mathbb{R}$-valued process W on a probability space $(\Omega, \mathcal{F}, P)$ with $W(0) = 0$, such that for each $\omega \in \Omega$ the sample path $\mathbb{R}_+ \ni t \to W(t, \omega) \in \mathbb{R}$ is continuous with probability one, and such that the law of W has finite dimensional distribution given by

$$P\left(\{W(t_1) \leq a^{(1)}, \ldots, W(t_k) \leq a^{(k)}\}\right) = \int_{-\infty}^{a^{(1)}} \frac{dx_1}{\sqrt{2\pi t_1}} e^{-\frac{x_1^2}{2t_1}} \cdot \qquad (A.2)$$

$$\cdot \int_{-\infty}^{a^{(2)}} \frac{dx_2}{\sqrt{2\pi(t_2-t_1)}} e^{-\frac{(x_2-x_1)^2}{2(t_2-t_1)}} \cdots \int_{-\infty}^{a^{(k)}} \frac{dx_k}{\sqrt{2\pi(t_k-t_{k-1})}} e^{-\frac{(x_k-x_{k-1})^2}{2(t_k-t_{k-1})}},$$

for $k \in \mathbb{N}$, $0 < t_1 < \cdots < t_k$, $a^{(i)} \in \mathbb{R}$, $i = 1, \ldots, k$.

As a consequence, the process W in (A.2), called *standard Brownian motion* or *Wiener process*, is an almost sure continuous Gaussian process with $W(0) = 0$ and $W(t)$ distributed according to

$$P(\{W(t) \leq a\}) = \int_{-\infty}^{a} \frac{dx}{\sqrt{2\pi t}} e^{-\frac{x^2}{2t}} =: \mathcal{N}(0,t)(a), \quad a \in \mathbb{R},$$

for $t > 0$, which has *independent increments*, i.e., $W(t)$ and $W(t+h) - W(t)$ are independent for $t, h > 0$. Moreover W is completely characterised by these properties. A standard Brownian motion in $\mathbb{R}^m$ is a process $W = (W_1, \ldots, W_m)$, where $W_1, \ldots, W_m$ are independent one-dimensional Brownian motions.

A.1.2 Martingales and Stopping Times

A *stopping time* τ with respect to a filtration $(\Omega, \mathcal{F})$ is an $\mathbb{R}_+$-valued random variable for which $\{\tau \leq t\} \in \mathcal{F}_t$ for every $t \geq 0$. Obviously, a stopping time τ is measurable with respect to the associated σ-algebra

$$\mathcal{F}_\tau := \{A \in \mathcal{F} : A \cap \{\tau \leq t\} \in \mathcal{F}_t, \quad t \geq 0\}.$$

A process M on the probability space $(\Omega, \mathcal{F}, P)$ is called a *martingale*, *submartingale*, or *supermartingale*, if for each $t \geq 0$, $E|M(t)| < \infty$ and, respectively,

$$E^{\mathcal{F}_s} M(t) = M(s), \qquad (A.3)$$
$$E^{\mathcal{F}_s} M(t) \geq M(s),$$
$$E^{\mathcal{F}_s} M(t) \leq M(s), \quad 0 \leq s \leq t \leq T_\infty.$$

The process M is said to be a *local martingale* if there exists a sequence of stopping times $(\tau_n)_{n \in \mathbb{N}}$ with $\tau_n \uparrow T_\infty$, such that for each n the stopped process $t \to M(t \wedge \tau_n)$ is a martingale in the sense of (A.3).

A *semi-martingale* X on $(\Omega, \mathcal{F}, P)$ is a right continuous process which can be written as

$$X = X(0) + M + A, \qquad (A.4)$$

where M is a local martingale with $M(0) = 0$, and A is a process of locally finite variation starting at zero. Hence, $A(0) = 0$, and for each fixed $t > 0$ we have

$$\sup_{\pi_t \in \mathcal{P}_{[0,t]}} \sum_{i=1}^{N} |A(t_i) - A(t_{i-1})| < \infty \quad a.s.,$$

where $\mathcal{P}_{[0,t]}$ denotes the set of partitions π_t on $[0,t]$, defined by the sequence $0 = t_0 < t_1 < \cdots < t_N = t$. For any semimartingale the decomposition (A.4), called *Doob–Meyer decomposition*, is unique.

REMARK A.1   The analogue definition of martingale, submartingale, and supermartingale for a discrete parameter processes $(M_n)_{n \in \mathbb{N}_0}$ is clear from (A.3). Moreover, if $(X_n)_{n \in \mathbb{N}}$ is a discrete parameter process adapted to a discrete filtration $(\mathcal{F}_n)_{n \in \mathbb{N}_0}$, it is easy to see that there exists a unique martingale M with $M_0 = 0$, and a process A with $A_0 = 0$ and A_n being $\mathcal{F}_{n-1}$ measurable for $n \geq 1$, such that $X = X_0 + M + A$. In particular, we have the discrete Doob–Meyer decomposition

$$X_n = X_0 + \underbrace{\sum_{k=0}^{n-1} \left(X_{k+1} - E^{\mathcal{F}_k} X_{k+1} \right)}_{M} + \underbrace{\sum_{k=0}^{n-1} \left(E^{\mathcal{F}_k} X_{k+1} - X_k \right)}_{A}, \quad n \geq 1.$$

Due to the measurability of A_n with respect to $\mathcal{F}_{n-1}$, A is called *predictable*. For continuous time processes, however, the notion of predictability is much more complicated. $\quad\Box$

A.1.3   Quadratic Variation and the Itô Stochastic Integral

Throughout this book we consider as integrators for stochastic integrals only semimartingales which are continuous, i.e., semimartingales which have almost sure continuous sample paths. We first introduce the notion of co-variation or bracket process and quadratic variation process for continuous semimartingales and then recall the Itô stochastic integral and his main properties.

A.1.3.1   The covariation process

Let M_1 and M_2 be continuous local martingales with $M_1(0) = M_2(0) = 0$. Then the *covariation process* or *bracket process* $\langle M_1, M_2 \rangle$ of M_1 and M_2 is defined as the finite variation part of the Doob–Meyer decomposition (A.4) of $M_1 M_2$; hence, $\langle M_1, M_2 \rangle(0) = 0$ and $M_1 M_2 - \langle M_1, M_2 \rangle$ is a local martingale. If X_1 and X_2 are semimartingales with Doob–Meyer decompositions $X_1 = X_1(0) + M_1 + A_1$ and $X_2 = X_2(0) + M_2 + A_2$ according to (A.4), then the covariation process $\langle X_1, X_2 \rangle$ of X_1 and X_2 is defined to be $\langle X_1, X_2 \rangle := \langle M_1, M_2 \rangle$, hence the covariation of the martingale parts of X_1

and X_2. For a continuous semi-martingale X, the *quadratic variation* $\langle X \rangle$ of X is defined to be $\langle X \rangle := \langle X, X \rangle$. Note that if $X = X(0) + M + A$ for a true martingale M, we have by Jensen's inequality, $E^{\mathcal{F}_s} M^2(t) \geq \left(E^{\mathcal{F}_s} M(t) \right)^2 = M^2(s)$, hence M^2 is a submartingale. From this it can be shown in general that the quadratic variation process $\langle M, M \rangle = \langle X, X \rangle$ is nondecreasing, and so in particular nonnegative.

It is easy to see that the covariation operator is symmetric and linear with respect to both arguments. Since moreover $\langle X \rangle = \langle X, X \rangle \geq 0$, we have the Cauchy–Schwartz inequality

$$|\langle X_1, X_2 \rangle| \leq \sqrt{\langle X_1, X_1 \rangle} \sqrt{\langle X_2, X_2 \rangle}. \qquad (A.5)$$

REMARK A.2   Since the martingale part of a finite variation process is zero, we have the following frequently used fact: If X and A are continuous semimartingales, and A has finite variation, then $\langle X, A \rangle = 0$.   ⬚

The terms covariation and quadratic variation are motivated by the following result. Let for $t > 0$, π_t be a partition on $[0, t]$, defined by the sequence, $0 = t_0 < t_1 < \cdots < t_N = t$, and let $\delta(\pi_t) := \max_{1 \leq i \leq N} \{t_i - t_{i-1}\}$. Then we have for two continuous semimartingales X_1 and X_2,

$$\langle X_1, X_2 \rangle(t) = \lim_{\delta(\pi_t) \to 0} \sum_{i=1}^{N} (X_1(t_i) - X_1(t_{i-1}))(X_2(t_i) - X_2(t_{i-1})),$$

where the limit is to be considered in probability.

REMARK A.3   The covariation process $\langle X_1, X_2 \rangle$, which is of finite variation, determines path-wise a measure on $\mathbb{R}_+$. In differential form this measure is usually denoted as $d\langle X_1, X_2 \rangle$, but also as $dX_1 \cdot dX_2$.   ⬚

A.1.3.2   The Itô stochastic integral

THEOREM A.1
*For a continuous local martingale M and a left continuous (predictable) process ϕ which satisfies $\int_0^t \phi^2(s) d\langle M \rangle(s) < \infty$ a.s. for all $t > 0$, there exists a unique **Itô stochastic integral** process I,*

$$I_{\phi, M}(t) := I(t) := \int_0^t \phi(s) dM(s), \qquad (A.6)$$

which satisfies the following properties,

(i) I is a local martingale with $I(0) = 0$.

(ii) For any continuous semimartingale X we have

$$\langle I, X \rangle = \int_0^t \phi(s) d\langle M, X \rangle(s),$$

where the latter integral is path-wise an ordinary Lebesgue–Stieltjes integral.

By taking $X = I$ in (ii), we have in particular

$$\langle I \rangle = \int_0^t \phi^2(s) d\langle M \rangle(s). \tag{A.7}$$

The Itô integral (A.6) can be generalized to any continuous semimartingale Y as integrator process in the following obvious way. If $Y = Y(0) + M + A$ is the Doob–Meyer decomposition of Y, define

$$I_{\phi,Y}(t) := \int_0^t \phi(s) dY(s) := \int_0^t \phi(s) dM(s) + \int_0^t \phi(s) dA(s), \tag{A.8}$$

provided that the last integral, which is path-wise an ordinary Lebesgue–Stieltjes integral since A is of finite variation, exists almost surely for all $t > 0$. It is easy to see by Remark A.2 that property (ii) in Theorem A.1 and thus (A.7) hold true for (A.8) as well. However, in general the local martingale property (i) is lost of course.

The interpretation of Theorem A.1 is as follows. For a partition $0 = t_0 < t_1 < \cdots < t_N < T_\infty \leq \infty$ and a sequence of random variables $(a_j)_{j \in \mathbb{N}_0}$, where a_j is $\mathcal{F}_j$ measurable for $j \geq 0$, consider the elementary predictable step process,

$$\phi(t, \omega) = \sum_{i=1}^N a_{i-1}(\omega) 1_{(t_{i-1}, t_i]}(t). \tag{A.9}$$

By straightforward technique it then follows that the process

$$I(t) := \sum_{i=1}^N a_{i-1} \left(M(t_i \wedge t) - M(t_{i-1} \wedge t) \right)$$

satisfies (i) and (ii) in Theorem A.1 and (A.7). For general integrands ϕ, the Itô integral (A.6) can be considered as a limit in probability of integrals due to a sequence of elementary step processes (A.9) converging to ϕ.

A.1.3.3 The Itô formula

For the Itô integral process we have the following substitution rule, called *Itô's lemma*.

LEMMA A.1

Let X be a continuous semimartingale in $\mathbb{R}^m$, i.e., $X = (X_1, \ldots, X_m)$ where X_i is a continuous semimartingale in $\mathbb{R}$ for $i = 1, \ldots, m$, and let $f \in C^2(\mathbb{R}^m)$, i.e., f has continuous partial derivatives of second order. Then, it holds

$$f(X(t)) = f(X(0)) + \sum_{i=1}^{m} \int_0^t \frac{\partial f}{\partial x_i}(X(s)) dX(s)$$

$$+ \frac{1}{2} \sum_{i,j=1}^{m} \int_0^t \frac{\partial^2 f}{\partial x_i \partial x_j}(X(s)) d\langle X_i, X_j \rangle(s). \tag{A.10}$$

By taking $f(x, y) = xy$ in Lemma A.1 we obtain the Itô integration by parts formula, also called the Itô product rule,

$$X_1(t) X_2(t) = X_1(0) X_2(0) + \int_0^t X_1(s) dX_2(s) + \int_0^t X_2(s) dX_1(s) + \langle X_1, X_2 \rangle(t). \tag{A.11}$$

A.1.3.4   Itô processes

The continuous stochastic processes involved in Chapters 1-6 are exclusively of Itô type. An Itô type process X in $\mathbb{R}^m$ is defined on a probability space $(\Omega, \mathcal{F}, P)$, where the filtration $\mathbb{F} = (\mathcal{F}_t)_{0 \leq t \leq T_\infty < \infty}$ is generated by a standard Brownian motion in $\mathbb{R}^m$ which satisfies the "usual conditions", and can be represented as a stochastic integral,

$$X_i(t) = X_i(0) + \sum_{i=1}^{n} \int_0^t \alpha_i(s) ds + \sum_{i=1}^{n} \sum_{j=1}^{m} \int_0^t \beta_{ij}(s) dW_j(s). \tag{A.12}$$

In (A.12) the processes α_i and β_{ij} are $\mathbb{F}$ predictable (continuous from the left is enough) and for all $t > 0$, $\int_0^t (|\alpha_i(s)| + |\beta_{ij}|) \, ds < \infty$ a.s. for $1 \leq i \leq n$, $1 \leq j \leq m$.

Of particular importance are positive valued Itô processes of the form

$$Y_i(t) = Y_i(0) \exp \left(-\frac{1}{2} \sum_{i=1}^{n} \int_0^t |\beta_i|^2(s) ds + \sum_{i=1}^{n} \int_0^t \alpha_i(s) ds \right.$$

$$\left. + \sum_{i=1}^{n} \sum_{j=1}^{m} \int_0^t \beta_{ij}(s) dW_j(s) \right), \tag{A.13}$$

where $Y_i(0) > 0$, $\beta_i := (\beta_{i1}, \ldots, \beta_{im})^\top$ and $|\beta_i|^2 = \beta_i^\top \beta_i$, for $1 \leq i \leq n$. By applying Itô's Lemma A.1 to (A.13) we obtain in differential form

$$\frac{dY_i(t)}{Y_i(t)} = \sum_{i=1}^{n} \alpha_i(t) dt + \sum_{i=1}^{n} \sum_{j=1}^{m} \beta_{ij}(t) dW_j(t). \tag{A.14}$$

The component solutions (A.13) of (A.14) are usually called Doléans exponentials and denoted as $Y_i = \mathcal{E}(X_i)$, where X_i is given by (A.12) for $1 \leq i \leq n$. The next lemma yields a convenient formula for the ratio of two Doléans exponentials.

LEMMA A.2
Let X and Y be given by

$$\frac{dX}{X} = \mu_X dt + \sigma_X^\top dW,$$
$$\frac{dY}{Y} = \mu_Y dt + \sigma_Y^\top dW$$

respectively. Then we have for X/Y,

$$\frac{d(X/Y)}{X/Y} = \left(\mu_X - \mu_Y - \sigma_Y^\top(\sigma_X - \sigma_Y)\right) dt + (\sigma_X - \sigma_Y)^\top dW. \qquad (A.15)$$

In terms of Doléans exponentials, Lemma A.2 reads

$$X = \mathcal{E}(P) \quad \wedge \quad Y = \mathcal{E}(Q) \quad \Rightarrow \quad \frac{X}{Y} = \mathcal{E}(P - Q - \langle P - Q, Q \rangle).$$

A.1.4 Regular Conditional Probability

The concept of regular conditional probability is in fact an extension of the elementary notion of conditional probability, $P(A|B) := P(A \cap B)/P(B)$ for a set B of positive measure, to conditioning sets of measure zero.

Let $\mathcal{G}$ be a sub σ-algebra of $\mathcal{F}$. Then a map

$$\Omega \times \mathcal{F} \ni (\omega, A) \rightarrow P(\omega, A)$$

is called a *regular conditional probability given $\mathcal{G}$*, if

(i) For fixed $\omega \in \Omega$, $P(\omega, \cdot)$ is a probability measure on $(\Omega, \mathcal{F})$;

(ii) For fixed $A \in \mathcal{F}$, the random variable $\omega \rightarrow P(\omega, A)$ is $\mathcal{G}$ measurable;

(iii) For any $\mathcal{F}$-measurable random variable Z it holds

$$[E^\mathcal{G} Z](\omega) = \int P(\omega, d\tilde{\omega}) Z(\tilde{\omega}) \qquad a.s.$$

According to a fundamental theorem (see, e.g., Ikeda and Watanabe (1981)) a regular conditional probability given $\mathcal{G} \subset \mathcal{F}$ exists and is unique, if the basic probability space $(\Omega, \mathcal{F}, P)$ is a *standard* probability space. For the definition of a standard probability space, see also Ikeda and Watanabe (1981). Without giving further details we just notice that all probability spaces considered in this book are standard.

A.2   Minimum Search Procedures

The objective functions encountered in the calibration procedures presented in Chapter 3 are yet smooth but may have more local minima. Therefore it is important to have a proper search routine for locating the global minimum of such functions. Although standard search routines such as Powel and Levenberg-Marguardt are basically local methods, they can be advantageously combined with global search methods based on minimization over random points or quasi-random points in the search space. The idea is as follows: Generate a number of starting points $s_1, \ldots, s_K$ in the search space and then start from each point a local minimum search, thus yielding a set of local minima $s_1^*, \ldots, s_K^*$. Then take as search result s_{j*}^*, where $f(s_{j*}^*) = \min_{1 \leq j \leq K} f(s_j^*)$ and f is the objective function. For the calibration experiments in Chapter 3, the Powel method is used as local search algorithm (for details see for instance Numerical Recipes in C), whereas quasi-random Halton points (see Section A.2.1) are used for the global searcher.

The advantage of quasi-random numbers over uniformly distributed Monte Carlo numbers is that their distribution in space is more uniform since the generation of these points in space proceeds self-avoiding in a sense. Due to this feature quasi-random points are also particularly suitable for high dimensional integration: If g is some smooth function on $[0,1]^d$ and $q_1, \ldots, q_N$ is a system of quasi-random numbers in $[0,1]^d$ (for instance Halton or Sobel points), then the accuracy of the estimation

$$\int_{[0,1]^d} g(x)dx \approx \frac{1}{N} \sum_{i=1}^{N} g(q_i)$$

is asymptotically of order of $O(\log^d N/N)$ which is essentially better than the order $O(1/\sqrt{N})$ error due to (pseudo) Monte Carlo random numbers (for large d the enumerator $\log^d N$ can become quite large however).

A.2.1   Halton Quasi-random Numbers

Halton points can be generated very easily. In one dimension the recipe for the generation of the j-th Halton point, $j \geq 1$, is as follows.

(i) Choose a prime p.

(ii) Write j as $j = a_k p^k + a_{k-1} p^{k-1} + \cdots + a_1 p + a_0$, for integers $0 \leq a_i < p$, $0 = 1, \ldots, k$, with $a_k \neq 0$.

(iii) Set $H_j^{(p)} = (a_0 p^k + a_1 p^{k-1} + \cdots + a_{k-1} p + a_k) p^{-(k+1)}$.

For arbitrary dimension d, $d \geq 1$, we take the first d prime numbers $p_1, \ldots, p_d$, and then the j-th d-dimensional Halton vector, $j \geq 1$, is given as

$$H_j = \left(H_j^{(p_1)}, \ldots, H_j^{(p_d)} \right).$$

Below we give C++ code for the generation of Halton points.[1] The main program in this code generates, as an example, the first ten Halton points in the five dimensional cube $[0, 1]^5$.

C++ code for generation of Halton points

```
class CHalton{
public:
CHalton();
    virtual ~CHalton();
    bool SetDimension(int dimension); //returns true, if no error occurred
    void GetHaltonVector(int i,double *x);
        // x points to the Storage, where dimension
        // double values between 0 and 1 will be stored.
        // i should be >=1.
    double Halton(int i,int base)const;
protected:
    double FastHalton(int i,int base)const;
private:
    int m_DIM;
    int *m_prime;
};

CHalton::CHalton(){
    m_DIM=0;
    m_prime=NULL;
}

CHalton::~CHalton(){
    if (m_prime!=NULL) delete [ ] m_prime;
}

bool CHalton::SetDimension(int dimension){
    if (m_prime!=NULL) delete [ ] m_prime;
    if (dimension≤0){
        m_DIM=0;
        m_prime=0;
        return dimension ==0;
```

---
[1] This code is written by O. Reiß.

```
    }
    m_prime=new int [dimension];
    if (m_prime==NULL){
        m_DIM=0;
        m_prime=0;
        return false;
    }
    m_DIM=dimension;
    int i,z=1;
    bool is_prim;
    for(dimension=0;dimension<m_DIM;dimension++){
        //Search the next prime
        do{
            z++;
            i=0;
            is_prim=true;
            while (is_prim && i<dimension){
                is_prim=is_prim && (z%m_prime[i]);
                i++;
            }
        }while (!is_prim);
        m_prime[dimension]=z;
    }
    return true;
}

void CHalton::GetHaltonVector(int i,double *x){
    int d;
    if (i<=0){
        for(d=0;d<m_DIM;d++) x[d]=0.0;
        return;
    }
    for(d=0;d<m_DIM;d++){
        x[d]=FastHalton(i,m_prime[d]);
    }
}

double CHalton::Halton(int i,int base)const{
    if (i<=0 || base <=1) return 0.0;
    return FastHalton(i,base);
}

double CHalton::FastHalton(int i,int base)const{
    const double y=1.0/base;
    double x=y;
```

```
    double erg=0.0;
    while(i!=0){
        erg+=x*(double)(i%base);
        i/=base;
        x*=y;
    }
    return erg;
}

void main(){
    CHalton hal;
    int DIM=5;
    int NumberOfHaltonPoints=10;
    int i,d;
    double *x;
    hal.SetDimension(DIM);
    x=new double[DIM];
    for(i=1;i<=NumberOfHaltonPoints;i++){
        hal.GetHaltonVector(i,x);
        for(d=0;d<DIM;d++){
            cout << x[d] << "\t";
        }
        cout << endl;
    }
    delete [ ] x;
}
```

A.3   Additional Proofs

A.3.1   Covariance of Two Black-Scholes Models

Consider two correlated Black-Scholes models driven by two independent Brownian motions W_1 and W_2,

$$\frac{dS_1}{S_1} = \mu_1 dt + \sigma_1 dW_1$$
$$\frac{dS_2}{S_2} = \mu_2 dt + \sigma_2(\rho dW_1 + \sqrt{1 - \rho^2}dW_2),$$

where $-1 \leq \rho \leq 1$. Then by multiplying the corresponding Doléans exponentials,

$$
\begin{aligned}
\frac{ES_1(T)S_2(T)}{S_1(0)S_2(0)} &= E \exp \left[\left(\mu_1 + \mu_2 - \frac{\sigma_1^2}{2} - \frac{\sigma_2^2}{2} \right) T \right. \\
&\quad \left. + (\sigma_1 + \sigma_2 \rho) W_1(T) + \sigma_2 \sqrt{1 - \rho^2} W_2(T) \right] \\
&= E \exp \left[\left(\mu_1 + \mu_2 - \frac{\sigma_1^2}{2} - \frac{\sigma_2^2}{2} \right) T + \sqrt{\sigma_1^2 + \sigma_2^2 + 2\rho\sigma_1\sigma_2} W_1(T) \right] \\
&= \exp \left[\mu_1 T + \mu_2 T + \rho \sigma_1 \sigma_2 T \right],
\end{aligned}
$$

where we use that $\exp[-a^2 T/2 + aW(T)]$ is a martingale. It then easily follows that

$$
\begin{aligned}
Cov(S_1(T), S_2(T)) &= ES_1(T)S_2(T) - (ES_1(T))(ES_2(T)) \\
&= S_1(0)S_2(0) \exp \left[\mu_1 T + \mu_2 T + \rho \sigma_1 \sigma_2 T \right] \\
&\quad - S_1(0)S_2(0) \exp \left[\mu_1 T + \mu_2 T \right] \\
&\approx S_1(0)S_2(0)\rho\sigma_1\sigma_2 T,
\end{aligned}
$$

when T is not too large.

A.3.2   Proof of Equality (5.26)

We here prove the equality (5.26). Let us write

$$
Y^{m+1}(i) - Y^m(i) = 1_{\tau_i^{(m+1)} = i} \left(Y^{m+1}(i) - Y^m(i) \right) + 1_{\tau_i^{(m+1)} > i} \left(Y^{m+1}(i) - Y^m(i) \right).
$$
$$(A.16)$$

Note that the first term in (A.16) is zero. Indeed, if $\tau_i^{(m+1)} = i$, then $Y^{m+1}(i) = E^{\mathcal{F}^{(i)}} Z(\tau_i^{(m+1)}) = Z^{(i)}$, and on the other hand, $Z^{(i)} \leq Y^m(i) \leq Y^{m+1}(i)$. Now we consider the second term in (A.16). If $\tau_i^{(m+1)} > i$, then $\tau_i^{(m+1)} = \tau_{i+1}^{(m+1)}$ and so $\tau_i^{(m+1)} > i \Rightarrow Y^{m+1}(i) = E^{\mathcal{F}^{(i)}} Z(\tau_i^{(m+1)}) = E^{\mathcal{F}^{(i)}} Z(\tau_{i+1}^{(m+1)}) = E^{\mathcal{F}^{(i)}} Y^{m+1}(i+1)$.

We thus obtain,

$$Y^{m+1}(i) - Y^m(i)$$

$$= 1_{\tau_i^{(m+1)} > i} \left(E^{\mathcal{F}^{(i)}} Y^{m+1}(i+1) - Y^m(i) \right)$$

$$= E^{\mathcal{F}^{(i)}} 1_{\tau_i^{(m+1)} > i} (Y^{m+1}(i+1) - Y^m(i+1)) + E^{\mathcal{F}^{(i)}} 1_{\tau_i^{(m+1)} > i} (Y^m(i+1) - Y^m(i))$$

$$= E^{\mathcal{F}^{(i)}} 1_{\tau_i^{(m+1)} > i} \left(E^{\mathcal{F}^{(i+1)}} 1_{\tau_{i+1}^{(m+1)} > i+1} (Y^{m+1}(i+2) - Y^m(i+2)) \right.$$

$$\left. + E^{\mathcal{F}^{(i+1)}} 1_{\tau_{i+1}^{(m+1)} > i+1} \left(Y^m(i+2) - Y^m(i+1) \right) \right)$$

$$+ E^{\mathcal{F}^{(i)}} 1_{\tau_i^{(m+1)} > i} \left(Y^m(i+1) - Y^m(i) \right)$$

$$= E^{\mathcal{F}^{(i)}} 1_{\tau_i^{(m+1)} > i+1} (Y^{m+1}(i+2) - Y^m(i+2)) +$$

$$+ E^{\mathcal{F}^{(i)}} 1_{\tau_i^{(m+1)} > i+1} \left(Y^m(i+2) - Y^m(i+1) \right)$$

$$+ E^{\mathcal{F}^{(i)}} 1_{\tau_i^{(m+1)} > i} \left(Y^m(i+1) - Y^m(i) \right).$$

Now, by induction from i to k and the fact that $Y^{m+1}(k) = Y^m(k)$, it follows that

$$Y^{m+1}(i) - Y^m(i) = E^{\mathcal{F}^{(i)}} \sum_{p=i}^{k-1} 1_{\tau_i^{(m+1)} > p} \left(Y^m(p+1) - Y^m(p) \right)$$

$$= E^{\mathcal{F}^{(i)}} \sum_{p=i}^{\tau_i^{(m+1)}-1} 1_{\tau_p^{(m)} = p} \left(E^{\mathcal{F}^{(p)}} Y^m(p+1) - Y^m(p) \right),$$

since $E^{\mathcal{F}^{(p)}} Y^m(p+1) = Y^m(p)$ on the set $\{\tau_p^{(m)} > p\}$.

A.3.3 Proof of Proposition 5.2

For all i, $0 \le i \le k$, and $m = 0, 1, 2 \ldots$, satisfying $m \ge k - i$ we have almost surely

$$Y^{m+1}(i) = Y^m(i),$$

i.e., the sequence becomes constant. We will prove this by backward induction over i. For $i = k$ the statement is true since for all $m \ge 0$,

$$Y^m(k) = Z(k).$$

Suppose now the statement is already proved for all j, $i + 1 \le j \le k$. By Theorem 5.1 we have for $0 \le j \le k$ and $m \ge 1$,

$$Y^m(j) \ge \tilde{Y}^m(j) = \max_{p:\, j \le p \le k} E^{\mathcal{F}^{(j)}} Z(\tau_p^{m-1})$$

$$\ge E^{\mathcal{F}^{(j)}} Z(\tau_{j+1}^{m-1}) = E^{\mathcal{F}^{(j)}} E^{\mathcal{F}^{(j+1)}} Z(\tau_{j+1}^{m-1})$$

$$= E^{\mathcal{F}^{(j)}} Y^{m-1}(j+1). \tag{A.17}$$

So, by Proposition 5.26 and the quasi-supermartingale property (A.17) we obtain for $m \geq 1$,

$$Y^{m+1}(i) - Y^m(i) = E^{\mathcal{F}^{(i)}} \sum_{p=i}^{\tau_i^{(m+1)}-1} E^{\mathcal{F}^{(p)}} Y^m(p+1) - Y^m(p)$$

$$\leq E^{\mathcal{F}^{(i)}} \sum_{p=i}^{\tau_i^{(m+1)}-1} E^{\mathcal{F}^{(p)}} Y^m(p+1) - E^{\mathcal{F}^{(p)}} Y^{m-1}(p+1)$$

$$\leq E^{\mathcal{F}^{(i)}} \sum_{p=i+1}^{k} Y^m(p) - Y^{m-1}(p), \tag{A.18}$$

where the terms in the sum are non-negative because $Y^m \geq Y^{m-1}$ due to Theorem 5.1 again. Suppose now that $m \geq k - i$. Hence, for all $p \geq i+1$, we have $m - 1 \geq k - p$. Then, by the induction hypothesis, we have $m \geq 1$ and $Y^{m-1}(p) = Y^m(p)$ for all $p \geq i+1$. Hence from (A.18) it follows that $Y^{m+1}(i) - Y^m(i) = 0$.

A.3.4 Proof of Proposition 5.3

$$\frac{V_0^{up^{up},K}}{B(0)} = E \sup_{1 \leq j \leq k} \left[\frac{C_{T_j}}{B(T_j)} - \sum_{q=1}^{j} \widetilde{Y}^{(q)} + \sum_{q=1}^{j} \frac{1}{K} \sum_{i=1}^{K} \xi_i^{(q)} \right]$$

$$= E \left[\frac{C_{T_{\widehat{j}_{\max}}}}{B(T_{\widehat{j}_{\max}})} - \sum_{q=1}^{\widehat{j}_{\max}} \widetilde{Y}^{(q)} + \sum_{q=1}^{\widehat{j}_{\max}} \frac{1}{K} \sum_{i=1}^{K} \xi_i^{(q)} \right]$$

$$= E \left(1 - 1_{[j_{\max} \neq \widehat{j}_{\max}]}\right) \left[\frac{C_{T_{j_{\max}}}}{B(T_{j_{\max}})} - \sum_{q=1}^{j_{\max}} \widetilde{Y}^{(q)} + \sum_{q=1}^{j_{\max}} \frac{1}{K} \sum_{i=1}^{K} \xi_i^{(q)} \right]$$

$$+ E 1_{[j_{\max} \neq \widehat{j}_{\max}]} \left[\frac{C_{T_{\widehat{j}_{\max}}}}{B(T_{\widehat{j}_{\max}})} - \sum_{q=1}^{\widehat{j}_{\max}} \widetilde{Y}^{(q)} + \sum_{q=1}^{\widehat{j}_{\max}} \frac{1}{K} \sum_{i=1}^{K} \xi_i^{(q)} \right]$$

$$= \frac{V_0^{up,K}}{B(0)} + O((P(j_{\max} \neq \widehat{j}_{\max}))^{1-1/q})$$

by Hölder's inequality, for any integer q. The latter assertion is due to the fact that for any q the q-th moment of both

$$\frac{C_{T_{\hat{j}_{\max}}}}{B(T_{\hat{j}_{\max}})} - \sum_{q=1}^{\hat{j}_{\max}} \widetilde{Y}^{(q)} + \sum_{q=1}^{\hat{j}_{\max}} \frac{1}{K} \sum_{i=1}^{K} \xi_i^{(q)}$$

and   (A.19)

$$\frac{C_{T_{j_{\max}}}}{B(T_{j_{\max}})} - \sum_{q=1}^{j_{\max}} \widetilde{Y}^{(q)} + \sum_{q=1}^{j_{\max}} \frac{1}{K} \sum_{i=1}^{K} \xi_i^{(q)}$$

exist and are uniformly bounded in K (we omit the proof). Then, since $\lim_{K \to \infty} P(j_{\max} \neq \hat{j}_{\max}) = 0$, the convergence for $K \to \infty$ of $V_0^{up^{up},K} \to V_0^{up,K}$ follows.

The convergence $V_0^{up_{low},K} \to V_0^{up,K}$ can be shown in a similar way.

References

[1] Andersen, L., A simple approach to the pricing of Bermudan swaptions in the multifactor LIBOR market model, *The Journal of Computational Finance*, 3 (Winter, 1999-2000), pp. 5–32.

[2] Andersen, L. and Andreasen, J., Volatility skews and extensions of the Libor market model, *Applied Mathematical Finance*, 7 (2000), pp. 1–32.

[3] Andersen, L. and Andreasen, J., Factor dependence of Bermudan swaption prices: Fact or fiction? *Journal of Financial Economics*, 62 (1) (2001), pp. 3–37.

[4] Andersen, L. and Broadie M., A primal-dual simulation algorithm for pricing multidimensional American options. Working paper (2001) .

[5] Belomestny, D. and Milstein, G.N., Monte Carlo evaluation of American options using consumption processes, Preprint No. 930, Weierstrass Institute Berlin (2004).

[6] Bender, C. and Schoenmakers, J., An iterative algorithm for multiple stopping: Convergence and stability, Preprint No. 991, Weierstrass Institute Berlin (2004).

[7] Berridge, S.J. and Schumacher, J.M., An irregular grid approach for pricing high-dimensional American options, Working paper (2004).

[8] Black, F., The pricing of commodity contracts, *Journal of Financial Economics*, 3 (1976), pp. 167–179.

[9] Brace, A., Gatarek, D. and Musiela, M., The market model of interest rate dynamics, *Mathematical Finance*, 7 (2) (1997), pp. 127–155.

[10] Brigo, D. and Mercurio, F., Joint calibration of the LIBOR market model to caps and swaptions volatilities, Working paper (2001).

[11] Brigo, D. and Mercurio, F., *Interest Rate Models,* Springer-Verlag, 2001a.

[12] Carmona, R. and Dayanik, S., Optimal multiple-stopping of linear diffusions and swing options. Working paper (2004).

[13] Carmona, R. and Touzi, N., Optimal multiple-stopping and valuation of swing options. Working paper (2004).

[14] Cont, R. and Tankov, P., *Financial Modelling with Jump Processes*, Chapman & Hall, 2003.

[15] Curnow, R.N. and Dunnett, C.W., The numerical evaluation of certain multivariate normal integrals, *Ann. Math. Statist.*, (33) (1962), pp. 571–579.

[16] Davis, M.H.A. and Karatzas, I., A deterministic approach to optimal stopping, *Probability, Statistics and Optimization, A Tribute to Peter Whittle*, F.P. Kelly (Ed.), John Wiley & Sons, New York & Chichester, 1994, pp. 455–466.

[17] Delbaen, F. and Schachermayer, W., A general version of the fundamental theorem of asset pricing, *Math. Ann.* 300 (1994), pp. 463–520.

[18] Duffie, D., *Dynamic Asset Pricing Theory*, Princeton University Press, Princeton, New Jersey, third edition, 2001.

[19] Duffie, D. and Glyn, P., Efficient Monte Carlo simulation of security prices. *The Annals of Applied Probability*, 5 (1995), pp. 897–905.

[20] Eberlein, E. and Özkan, F., The Lévy Libor Model, FDM Preprint, A-L University Freiburg (2004).

[21] Elliot, R.J. and Kopp, P.E., *Mathematics of Financial Markets*, Springer-Verlag, New York-Berlin-Heidelberg, 1999.

[22] Fisher, M., Nychka, D. and Zervos, D., Fitting the term structure of interest rates with smoothing splines, Finance and Economics Discussion Series, Federal Reserve Board, Washington, D.C. (1995).

[23] Glasserman, P., *Monte Carlo Methods in Financial Engineering*. Springer-Verlag, Berlin-Heidelberg-New York, 2003.

[24] Glasserman, P. and Kou, S., The term structure of simple forward rates with jump risk. *Mathematical Finance*, 13 (2003), pp. 277–300.

[25] Glasserman, P. and Merener, N., Cap and swaption approximations in LIBOR market models with jumps, Working paper (2002).

[26] Harisson, M.J. and Pliska, S.R., Martingales and stochastic integrals in the theory of continuous trading, *Stochastic Processes and Their Applications*, 11 (1981), pp. 215–260.

[27] Harisson, M.J. and Pliska, S.R., A stochastic calculus model of continuous trading: complete markets, *Stochastic Processes and Their Applications*, 15 (1983), pp. 313–316.

[28] Haugh, M.B. and Kogan L., Pricing American options: A duality approach. *Operations Research*, 52 (2) (2004), pp. 258–270.

[29] Heath, D., Jarrow, R. and Morton, A., Bond pricing and the term structure of interest rates: A new methodology for contingent claim valuation, *Econometrica,* 60, (1) (1992), pp. 77–105.

[30] Howard R., *Dynamic Programming and Markov Processes.* MIT Press, Cambridge, Massachusetts, 1960.

[31] Hull, J. and White, A., Forward rate volatilities, swap rate volatilities, and the implementation of the Libor market model, *Journal of Fixed Income,* 10 2 (2000), pp. 46–62.

[32] Ikeda, N. and Watanabe, S., *Stochastic Differential Equations and Diffusion Processes,* North-Holland, Amsterdam, 1981.

[33] Jäckel, P. and Rebonato, R., The link between caplet and swaption volatilities in a Brace/Gatarek/Musiela/Jamshidian framework: Approximate solutions and empirical evidence, *Journal of Computational Finance,* 6 (4) (2003), pp. 41–59.

[34] Jamshidian, F., LIBOR and swap market models and measures. *Finance and Stochastics,* 1 (4) (1997), pp. 293-330.

[35] Jamshidian, F., LIBOR market model with semimartingales, in *Option Pricing, Interest Rates and Risk Management,* E. Jouini, J. Cvitanic, and Marek Musiela (Eds.), Cambridge Universty Press, 2001.

[36] Jamshidian, F., Minimax optimality of Bermudan claims: An arbitrage argument approach without probability theory, Working paper (2003).

[37] Jamshidian, F., Numeraire-invariant option pricing & american, bermudan, and trigger stream rollover, Working paper (2004).

[38] Joshi, M.S. and Theis, J., Bounding Bermudan swaptions in a swap-rate market model, *Quantitative Finance,* 2 (5) (2002), pp. 371–361.

[39] Kallenberg, O., *Foundations of Modern Probability,* 2nd ed. Springer-Verlag, 2002.

[40] Karatzas, I. and Shreve, S.E., *Brownian Motion and Stochastic Calculus,* 2nd ed. Springer-Verlag, 1991.

[41] Kloeden, P.E. and Platen, E., *Numerical Solution of Stochastic Differential Equations,* Springer-Verlag, Series Applications of Mathematics, Nr. 23, 1992.

[42] Kolodko, A. and Schoenmakers, J., An efficient dual Monte Carlo upper bound for Bermudan style derivative, Preprint No. 877, Weierstrass Institute Berlin (2003).

[43] Kolodko A. and Schoenmakers, J., Iterative construction of the optimal Bermudan stopping time, Preprint No. 926, Weierstrass Institute Berlin

(2004), Proc. 2nd IASTED Fin. Eng. Appl., Cambridge MA, pp. 230-238.

[44] Kolodko, A. and Schoenmakers, J., Upper bounds for Bermudan style derivatives, *Monte Carlo Methods and Appl.*, 10 (3-4) (2004a), pp. 331-343.

[45] Kurbanmuradov, O., Sabelfeld, K. and Schoenmakers, J., Lognormal approximations to Libor market models, *Journal of Computational Finance*, 6 (Fall, 2002), pp. 69-100.

[46] Linton, O., Mammen, E., Nielsen, J. and Tanggaard, C., Estimating yield curves by kernel smoothing methods. *Journal of Econometrics*, 105 (1) (2001), pp. 185-223.

[47] Malherbe, E., Correlation analysis in the LIBOR and swap market model, *Int. J. of Theoretical and Applied Finance*, 5 (4) (2002), pp. 401-426.

[48] Meinshausen, N. and Hambly, B. M., Monte Carlo methods for the valuation of multiple-exercise options, *Math. Finance*, 14 (2004), 557-583.

[49] Milstein, G.N., *Numerical Integration of Stochastic Differential Equations*, Kluwer Academic Publishers (Engl. transl. from Russian 1988) 1995.

[50] Milstein, G.N., Reiß, O. and Schoenmakers, J., A new Monte Carlo method for American options, *Int. J. of Theoretical and Applied Finance*, 7, 5 (2004), pp. 591-614.

[51] Milstein, G.N. and Schoenmakers, J.G.M., Monte Carlo construction of hedging strategies against multi-asset European claims, *Stochastics and Stochastics Reports*, 73, (1-2), pp. 125-157, (2002).

[52] Milstein, G.N. and Tretyakov, M.V., *Stochastic Numerics for Mathematical Physics*, Springer-Verlag, Berlin Heidelberg New York, 2004.

[53] Miltersen, K.R., Sandmann, K. and Sondermann, D., Closed-form solutions for term structure derivatives with lognormal interest rates, *J. Finance*, 52 (1997), pp. 409-430.

[54] Musiela, M. and Rutkowski, M., Continuous-time term structure models: Forward measure approach, *Finance and Stochastics*, 1 (1997), pp. 261-292.

[55] Press, W.H., Teukolsky, S.A., Vetterling, W. T. and Flannery, B.P., *Numerical Recipes in C: The Art of Scientific Computing*, 2nd ed., Cambridge University Press, Cambridge, 1992.

[56] Puterman M., *Markov Decision Processes*, Wiley, New York, 1994.

[57] Rapisarda, F., Brigo, D. and Mercurio, F., Parameterizing correlations: A geometric interpretation, Product and Business Development Group, Banca IMI, Working paper (2002).

[58] Rebonato, R., *Interest-rate Option Models,* John Wiley & Sons, Chichester, 1996.

[59] Rebonato, R., *Volatility and Correlation,* John Wiley & Sons, 1999.

[60] Rebonato, R., On the simultaneous calibration of multifactor lognormal interest rate models to Black volatilities and to the correlation matrix, *Journal of Computational Finance,* 2 (4) (1999a), pp. 5–27.

[61] Reiß, O., Schoenmakers, J. and Schweizer, M., Endogenous interest rate dynamics in asset markets, Preprint No. 652, Weierstrass Institute Berlin (2001).

[62] Reiß, O., Schoenmakers, J. and Schweizer, M., From structural assumptions to a link between assets and interest rates, Preprint DFG center MATHEON Berlin (2004).

[63] Revuz, D. and Yor, M., *Continuous Martingales and Brownian Motion,* Springer-Verlag, Berlin-Heidelberg-New York, 1998.

[64] Rogers, L.C.G., Monte Carlo valuation of American options, *Mathematical Finance,* 12 (2001), pp. 271–286.

[65] Rousseeuw, P.J., Molenberghs, G., Transformation of non positive semi-definite correlation matrices, *Commun. Statist. - Theory Meth.,* 22 (1993), pp. 965–984.

[66] Sabelfeld, K.K., *Monte Carlo Methods in Boundary Value Problems,* Springer-Verlag, (1991).

[67] Schoenmakers, J. and Coffey, B., LIBOR rate models, related derivatives and model calibration, QMF 1998 Sydney, Preprint No. 480, Weierstrass Institute Berlin (1999).

[68] Schoenmakers, J. and Coffey, B., Stable implied calibration of a multi-factor LIBOR model via a semi-parametric correlation structure, Preprint No. 611, Weierstrass Institute Berlin (2000).

[69] Schoenmakers, J. and Coffey, B., Systematic generation of parametric correlation structures for the LIBOR market model, *International Journal of Theoretical and Applied Finance,* 6 (5) (2003), pp. 507–519.

[70] Schoenmakers, J.G.M. and Heemink, A.W., Fast valuation of financial derivatives, *Journal of Computational Finance,* 1 (1) (1997), pp. 47–62.

[71] Schoenmakers, J., Calibration of LIBOR models to caps and swaptions: A way around intrinsic instabilities via parsimonious structures and a

collateral market criterion, Preprint No. 740, Weierstrass Institute Berlin (2002).

[72] Schroder, M., Computing the constant elasticity of variance option pricing formula, *Journal of Finance*, 44 (1) (1989), pp. 211–218.

Index